5th International Seminar on Ocean and Coastal Engineering, Environmental and Natural Disaster Management (ISOCEEN 2017)

MATEC Web of Conferences Volume 177 (2018)

Surabaya, Indonesia
8 November 2017

Editors:

Suntoyo **Agro Wisudawan**

ISBN: 978-1-5108-6600-3

Printed from e-media with permission by:

Curran Associates, Inc.
57 Morehouse Lane
Red Hook, NY 12571

Some format issues inherent in the e-media version may also appear in this print version.

Printed by Curran Associates, Inc. (2018)

For additional information, please contact EDP Sciences – Web of Conferences at the address below.

EDP Sciences – Web of Conferences
17, Avenue du Hoggar
Parc d'Activité de Courtabœuf
BP 112
F-91944 Les Ulis Cedex A
France

Phone: +33 (0) 1 69 18 75 75
Fax: +33 (0) 1 69 28 84 91

contact-edps@webofconferences.org

Additional copies of this publication are available from:

Curran Associates, Inc.
57 Morehouse Lane
Red Hook, NY 12571 USA
Phone: 845-758-0400
Fax: 845-758-2633
Email: curran@proceedings.com
Web: www.proceedings.com

TABLE OF CONTENTS

Preface

International Seminar on Ocean and Coastal Engineering, Environmental and Natural Disaster Management (ISOCEEN) is an annual internasional conference event held by cooperation among Institut Teknologi Sepuluh Nopember (ITS) especially Department of Ocean Engineering, Tohoku University, Japan, HZ University of Applied Sciences, Netherlands and The Coral Triangle Initiative on Coral Reefs, Fisheries and Food Security (CTI-CFF). This forum was more focused on creating and expanding professional network to foster the relationship between the University, industry, business and communities across the country. The Organizing Committee worked many years for preparing of this seminar and their efforts resulted to full success. Based on the cooperation, there are some conference that we already held before, as follows :

- Japan – Indonesia Workshop on Estuary and Climate Change (JIWECC), August 8th -10th Surabaya, Indonesia
- The 1st International Seminar on Ocean and Coastal Engineering, Environmental and Natural Disaster Management (ISOCEEN), November 11, 2013,. Surabaya, Indonesia.
- The 2nd International Seminar on Ocean and Coastal Engineering, Environmental and Natural Disaster Management (ISOCEEN), December 2, 2014, Surabaya, Indonesia.
- The 3rd International Seminar on Ocean and Coastal Engineering, Environmental and Natural Disaster Management (ISOCEEN), December 10, 2015, Surabaya, Indonesia.
- The 4th International Seminar on Ocean and Coastal Engineering, Environmental and Natural Disaster Management (ISOCEEN), December 13, 2016, Surabaya, Indonesia.
- The 5th International Seminar on Ocean and Coastal Engineering, Environmental and Natural Disaster Management (ISOCEEN), November 8, 2017, Surabaya, Indonesia

ISOCEEN 2017 was aimed on providing a discussion, for exchanging of knowledge, researches and of recent solutions for many researchers and experts in the field of Ocean, Offshore, Coastal engineering, Environmental and Disaster Management. The results of discussions are presented in this book. We hope this book will be interesting and usefull for many engineers and scientists whose activity related with researches of ocean and development of coastal areas.

Finally, on behalf of the organizing committee of the seminar, sincere appreciation is expressed to all authors who contribute to our seminar. Special thanks are also due to all keynote speakers, invited talks and chairpersons for the efforts in preparing the manuscripts and managing the sessions, respectively.

Suntoyo, PhD
Chairman of Organizing Committee

Committees

Advisory Committee:

- Rector of ITS Surabaya, Indonesia
- Dean of Marine Technology Faculty - ITS, Indonesia
- Head of Ocean Engineering Department, Faculty of Marine Technology - ITS, Indonesia

Organizing Committee:

- Suntoyo, PhD (Chairman)
- Silvianita, PhD (Secretary)
- Yoyok Setyo Hadiwidodo, PhD
- Dr. Hasan Ikhwani, MSc
- Dr. Wahyudi
- Dr. Eng. Rudi Walujo Prastianto
- Dr. Eng. Muhammad Zikra
- Agro Wisudawan, MT
- Dirta Marina Chamelia, MT

International Scientific Committee:

- Prof. Hitoshi Tanaka, Dr. Eng. (Tohoku University, Japan)
- Prof. Widi Agoes Pratikto, PhD (ITS, Indonesia)
- Prof. Daniel M. Rosyid, PhD (ITS, Indonesia)
- Prof. Eko Budi Djatmiko, PhD (ITS, Indonesia)
- Prof. Mukhtasor, PhD (ITS, Indonesia)
- Suntoyo, PhD (ITS, Indonesia)
- Makoto Umeda, Dr. Eng. (Tohoku University, Japan)
- Dr. Joost Stronkhorst (HZ University of Applied Sciences, Netherlands)

- Ir. Liliane Geerling R.T.D. (HZ Universityof Applied Sciences, Netherlands)
- Ahmad Sana, PhD (Sultan Qaboos University, Oman)
- Priyantha Ranjan, PhD (Curtin University, Australia)
- Dr. Eng. Totok Suprijo (ITB, Indonesia)
- Dr. Eng. Purwanto Bekti Santoso (Unsoed, Indonesia)
- Dr. Eng. Sitang Pilailar (Kasetsart University, Thailand)
- Dr. Nguyen Trung Viet (Water Resource University, Vietnam)
- Dr. Taufiqur Rachman (Hasanuddin University, Indonesia

Opening Speech by Chairman of ISOCEEN 2017 for International Seminar on Ocean and Coastal Engineering, Environmental and Natural Disaster Management, on Tuesday, 08 November, 2017 at 08.45 am at Swiss Bellin Manyar Hotel, Surabaya

The first that we respect, Rector of ITS, Prof. Joni Hermana, Head of Ocean Engineering Department, Dr.Eng. Rudi Walujo P, Keynote Speaker, Dr. Alvaro Semedo, Dr. Ali Dastgheib, from Unesco-IHE, Netherlands, Invited talks Dr. Mohd Faris Khamidi, University of Reading Malaysia, Liliane Geerling, R.TD, HZ University of Applied Sciences-Delta Academy, Netherlands, Lecturers within ITS, Contributing Authors, Seminar delegates , experts, scholar, all student, Ladies and gentlemen.

Assalamu alaikum Wr. Wb., Good morning! Presence of Allah SWT 's grace that has given favors and gifts in this morning. Welcome to this **The 5th International Seminar on Ocean and Coastal Engineering, Environmental and Natural Disaster Management (ISOCEEN 2017)**. I am pleased to join you here today.

We are delighted to announce that Institut Teknologi Sepuluh Nopember (ITS) especially Department of Ocean Engineering and Tohoku University have collaboration since 1997. In which, in the year 2006, a Memorandum of Understanding (MoU) has been signed between Tohoku University, Japan and Institut Teknologi Sepuluh Nopember to promote further collaboration between two institutions and and it has been extended until 20 December 2016. Many activities that have performed well in term of collaborative research, faculty and student exchange, joint educational program, seminars, workshops etc, especially for the seminar and workshop has been done several times before, namely; Japan-Indonesia Estuary Workshop, August 22nd, 2005 Jakarta, Japan – Indonesia Workshop on Estuary and Climate Change (JIWECC), August 8th -10th Surabaya, Indonesia. However the previous seminar activities organized not regularly therefore Since 2013 held an annual seminar entitled The first International Seminar on Ocean, Coastal Engineering, Environmental and Natural Disaster Management (ISOCEEN), November 11, 2013, Surabaya Indonesia.The 2nd International Seminar on Ocean and Coastal Engineering, Environmental and Natural Disaster Management (ISOCEEN), December 2, 2014. In which the selected papers have been published to the International Journal, Procedia Earth and Planetary Science. http://www.sciencedirect.com/science/journal/18785220/14

The 3rd ISOCEEN has been organized at 10 December 2015. The selected papers have been published to Trans Tech Publication on Applied Mechanics and Material Volume 862 in Special Edition Ocean Science and Coastal Engineering indexed in ISI Thomshon Rheuter.

Moreover, a Memorandum of Understanding (MoU) has been signed on July 29, 2015 between Institut Teknologi Sepuluh Nopember and HZ University of Applied Sciences, Netherlands to promote the development of academic scientific and cultural exchange, and of teaching and research activities, especially with Department of Ocean Engineering, ITS. And in 2016-2019

Our Department also have grants from Nuffic-Niche Project to increase collaboration among ITS-ITB and TU Delft Consorsium, Netherlands.

The 4th ISOCEEN has been organized at 13 December 2016. The selected papers have been published to to "*IOP Conference Series: Earth and Environmental Science*" indexed in Scopus. http://iopscience.iop.org/issue/1755-1315/135/1

With the increasing amount of cooperation between Indonesia, Japan, Netherlands and another country then this forum is more focused on creating and expanding professional network to foster the relationship between the University, industry, business and communities across the country. Therefore it is necessary to hold **The 5th International Seminar on Ocean and Coastal Engineering, Environmental and Natural Disaster Management (ISOCEEN) 2017** in which the selected papers presented in the seminar will be published to "*MATEC Web of Conferences*" indexed in Scopus.

Finaly, on behalf of the organizing committee of the seminar, sincere appreciation is expressed to all speakers contributing to our seminar. Special thanks are also due to all keynote speakers and all committee for the efforts in preparing the manuscripts and managing the sessions, respectively.

Wish the seminar a complete success.

Thank you !

ISOCEEN 2017

The 5th International Seminar on Ocean and Coastal Engineering, Environmental and Natural Disaster Management (ISOCEEN) 2017

8TH NOVEMBER 2017

SUBMITTED PAPER

Secretariat

- Department of Ocean Engineering, Faculty of Marine Technology Institut Teknologi Sepuluh Nopember (ITS)
  Kampus ITS – Keputih – Sukolilo, Surabaya (60111) -Indonesia
  Telp : +62 31 5928105 | Fax : +62 31 5928105
  Email : iso-ceen@its.ac.id / iso-ceen@gmail.com
  CP : +62 82232354708 (Silvianita, PhD) . +6285856150310 (Agro Wisudawan, MT)

Seminar Themes

- Hydrodynamics and Morphodynamics
- Ocean Engineering and Offshore Technology
- Port and Coastal Engineering
- Coastal, Delta, River and Estuary Management and Its Environmental
- Natural Disaster Management
- Sea Transportation and Ship Technology
- Ocean renewable Energy and Development

Keynote Speakers and Invited Talks:

1. Prof. Dr.Eng Hitoshi Tanaka, Vice-president, Japan Society of Civil Engineering (JSCE), Tohoku University, Japan
2. Prof. Dr. Julie Pietrzak, Professor Physical Oceanography, TU Delft, Netherlands. (just confirmed via email)
3. Prof. Widi Agoes Pratikto, Ph.D, Executive Director, CTI CFF
4. Assoc. Prof. Alvaro Semedo, Ph.D, Unesco-IHE, Netherlands
5. Ir. Liliane Geerling, R.TD, HZ University of Applied Sciences-Delta Academy, Netherlands

Introduction

We are delighted to announce that Institut Teknologi Sepuluh Nopember (ITS) especially Department of Ocean Engineering and Tohoku University have collaboration since 1997. In which, in the year 2006, a Memorandum of Understanding (MoU) has been signed between Tohoku University, Japan and Institut Teknologi Sepuluh Nopember to promote further collaboration between two institutions and and it has been extended until 20 December 2016. Many activities that have performed well in term of collaborative research, faculty and student exchange, joint educational program, seminars, workshops etc, especially for the seminar and workshop has been done several times before, namely; Japan-Indonesia Estuary Workshop, August 22nd, 2005 Jakarta, Japan – Indonesia Workshop on Estuary and Climate Change (JIWECC), August 8th -10th Surabaya, Indonesia and The first International Seminar on Ocean, Coastal Engineering, Environmental and Natural Disaster Management (ISOCEEN), November 11, 2013, Surabaya Indonesia. The 2nd International Seminar on Ocean and Coastal Engineering, Environmental and Natural Disaster Management was held on December 2, 2014. in which the selected papers have been published to the International Journal, Procedia Earth and Planetary Scence. http://www.sciencedirect.com/science/journal/18785220/14.

The 3rd International Seminar on Ocean and Coastal Engineering, Environmental and Natural Disaster Management was held on December 10, 2015. Published to Applied Mechanics and Material, https://www.scientific.net/AMM.862

The 4th International Seminar on Ocean and Coastal Engineering, Environmental and Natural Disaster Management was held on December 13, 2016, the selected paper was processed to publish into IOP Conference Series – Earth and Environmental Sciences *indexed in Scopus* .

The 5th International Seminar on Ocean and Coastal Engineering, Environmental and Natural Disaster Management will be held on November 8, 2017. The selected paper will be published into: IOP Conference Series – Earth and Environmental Sciences *indexed in Scopus.*

Call For Papers

The technical program of the Conference will consist of keynote addresses, contributed papers and special sessions with invited speakers. Authors are invited to submit an extended abstracts (short papers) of 2-3 pages, including some results and/or graphics (ISOCEEN-2017-Extended AbstractTemplate). The extended abstracts will be reviewed and selected by an International Technical Committe for oral presentation. Prospective authors are requested to submit an extended abstracts to the Seminar Secretariat. It may be submitted either online through conference website or via regular mail. After the seminar, the full paper must be submitted and will be reviewed to be aligned with the requirements published in MATEC Web of Conference *indexed in Scopus*.

IMPORTATNT DATES

Deadline for Extended abstracts submission	21 Oktober 2017
Notification of Acceptance	24 Oktober 2017
Deadline for Authors'Registration	30 Oktober 2017
Seminar Date	8 November 2017
Deadline for Full Paper submission	30 December 2017
Review process	1-30 January 2018
Revision process	1-28 February2018
Final checking and editing by Editors	30 May 2018
Deliver the Final Manuscript to the Publisher	15 June 2018

REGISTRATION FEE

	Foreign Delegates	Indonesian Delegates
Delegates	US $ 75	IDR 600,000
Contributing Author	US $ 400	IDR 3,000,000
Accompanying Person	US $ 75	IDR 600,000
Extra One Paper	US $ 200	IDR 2,000,000

TOHOKU UNIVERSITY

ep nuffic Internationalising education

ISOCEEN 2017

HZ UNIVERSITY OF APPLIED SCIENCES

edp sciences

Web of Conferences

Statement of Peer review

In submitting conference proceedings to *Web of Conferences*, the editors of the proceedings certify to the Publisher that

1. They adhere to its **Policy on Publishing Integrity** in order to safeguard good scientific practice in publishing.
2. All articles have been subjected to peer review administered by the proceedings editors.
3. Reviews have been conducted by expert referees, who have been requested to provide unbiased and constructive comments aimed, whenever possible, at improving the work.
4. Proceedings editors have taken all reasonable steps to ensure the quality of the materials they publish and their decision to accept or reject a paper for publication has been based only on the merits of the work and the relevance to the journal.

Title, date and place of the conference

The 5th International Seminar on Ocean and Coastal Engineering, Environmental and Natural Disaster Management (ISOCEEN) 2017, November 8th, 2017 in Surabaya, Indonesia

Proceedings editor(s): Suntoyo, PhD and Agro Wisudawan, MT

Date and editor's signature 15 June 2018

MATEC Web of Conferences **177**, 01001 (2018) https://doi.org/10.1051/matecconf/201817701001
ISOCEEN 2017

Intact stability and seakeeping characteristics of autonomous surface vehicle (asv) using swath hullform to support bathymetry survey activities on the coastal area

Ahmad Fauzan Zakki[1,*], *Aris* Triwiyatno[2], *Bandi* Sasmito[3], and *Muhammad H.* Nubly[1]

[1]Naval Architecture Department, Engineering Faculty, Diponegoro University, Indonesia
[2]Electrical Engineering Department, Engineering Faculty, Diponegoro University, Indonesia
[3]Geodetical Engineering Department, Engineering Faculty, Diponegoro University, Indonesia

Abstract. The information is an important thing to search a new resource. In this decade, with leading-edge technology is easier to find any information about our earth. Thus, that is must be explored by the ships. The exploration works usually take a long time, and for convenience, the unmanned technology is much promised. In this research, we focused on the autonomous surface vehicle for bathymetry survey. To design the ASV, we must consider the reliability, efficiency, and safety. To reach those three components, the analysis of resistance, stability and wave response are necessary. Therefore, the hull form design is focused on Small Waterplane Area Twin Hull (SWATH) hull form. According to the numerical analysis, the proposed SWATH hull form of ASV is suitable for bathymetry activities. It can be proved by the stability performance which the maximum value of GZ is at 26.40. The resistance calculation result is indicated that the drag of coefficient of this hull form is about 0.013.

1 Introduction

At the previous time, bathymetry survey was performed by drop down the cables or ropes through the seabed with the ship. However, that technique only limited by once survey per point so that is not efficient. At modern ages, the bathymetry survey can be performed by echo sounding which is attached to the ship side and then radiated the waves. The traveling time of radiated wave from the surface is reflected by seabed and then received. This method was used to calculate the deep of water. As it develops of autonomous surface vehicle (ASV) technology, the role of the ship has been changed by an autonomous vehicle. The marine vehicle is the first vehicle which is using autopilot since 1920's. However, the autopilot which is used on the modern ship is not significantly different with the previous technology, which might not be adapted to the dynamic of the sea environment. The advanced control technology for Unmanned Aerial Vehicle (UAV) and Autonomous Land Vehicle (ALV) is not affected by the marine vehicle. The marine environment is having the difficult characteristic to predict and need more power consumption for propeller system. The prediction of seakeeping performance is necessary, that is for predict the resistance of ship due to responses of waves, [1]. For long distance operational such as oceanography exploration and survey, an ASV which is equipped with power onboard and sophisticated control of navigation are more promising by less human intervention to operate. The robotics technology for the marine vehicle was bringing through to access deep water survey, resource mapping, and military. This ability is precious for an archipelago state which has marine resources.

* Corresponding author: ahmadfzakki@undip.ac.id

Several articles have been reviewed to support the ASV hull form design using SWATH type hull form. These articles relate to the study of SWATH hull form characteristics and designs and the development of motion controls on ASV. In the development of the SWATH hull form, Gregor Schellenberger, [1] studied the SWATH hull form design method. Parametric studies were performed to obtain optimal SWATH hull form. Some SWATH hull types such as circular shape, Golf Club, and Elephant foot was proposed. Twin Strut or Single strut are also reviewed. The application of the optiSWATH tool is used to obtain the optimum design of the SWATH hull form, [2]. A.E. Baitis et.al presented the seakeeping analysis by comparing monohull, catamaran, and SWATH, [3]. The development study of ASV technology, MIT Sea Grant College Program was developing an Autonomous Surface Craft as known as ARTEMIS, this ASV is a trawler replica, C. Goudey et. al was performed an ASV for the necessity of fish tracking. J.Manley and T.Vaneck were developing a continuous technology of ARTEMIS which is expanded for hydrography survey necessity, [4]. US Navy was shown their interest of ASV by focused on ASV function as transmission equipment for communication support, [5]. Another study, J.Curcio et. al has performed the study of ASV for moving baseline navigation, [6]. The innovative hull form use by using multihull was developed, [7],[8]. Caccia M. was developing SESAMO for sampling on sea surface using semi-submersible hull. H. Ferreira also was developing an ASV using catamaran hull form which named as Swordfish and as known as ROAZ II, [9]. S. Phillips was developing an ASV using semi-submersible hull form for hydrography survey, [10]. W. Sager has performed a study by using the sonar on ASV which is used for hydrography survey, [11]. R. Hine and P. McGillivary have performed a study about wave generator used for the power source of ASV [12]. M. Benjamin was developing a system for ASV collision avoidance, [13].

Based on the reviewed articles above, this research is intended to develop a design for ASV using SWATH hull form to support bathymetry survey. The advantages of SWATH hull form can be expected to obtain the ASV's seakeeping.

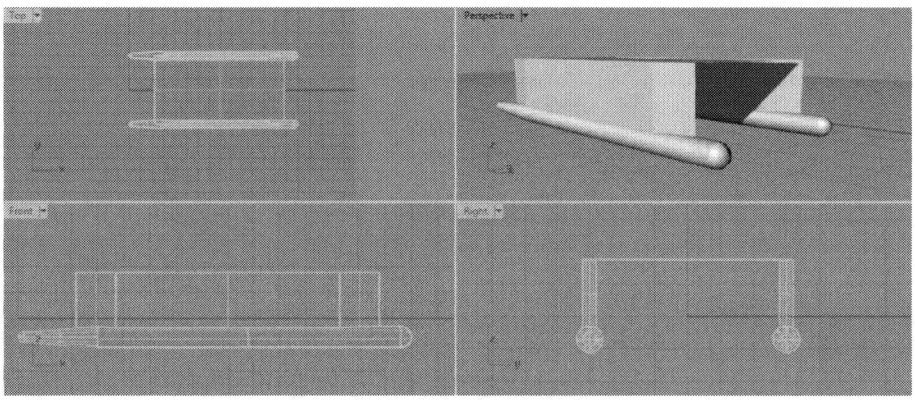

Fig. 1. The 3D model of SWATH Autonomous Surface Vehicle

2 General designs and concept

To obtain the principal dimension could be compared the length, breadth, height, draught, and displacement of parent ship. The developed SWATH hull form of the autonomous surface vehicle is shown in the **Fig. 1** and the principal dimension can be seen in **Table 1**.

Table 1. The principal dimensions of ASV

Principal Dimensions	
Length	1500 mm
Breadth	680 mm
Height	280 mm
Draught	110 mm
Displacement (Δ)	41.19 Kg

3 Result and discussions

3.1 Intact stability characteristics

Stability calculations were necessary to obtain the GZ moment of the ASV. To calculate it, we need an assumption of the ASV load cases. In each station, must defining the lightship and any stuff which is ASV carried. Distributing the load for each station and finding the point mass. For stability criteria, we are using IMO criteria for general merchant ship (IMO A.749), in intact conditions.

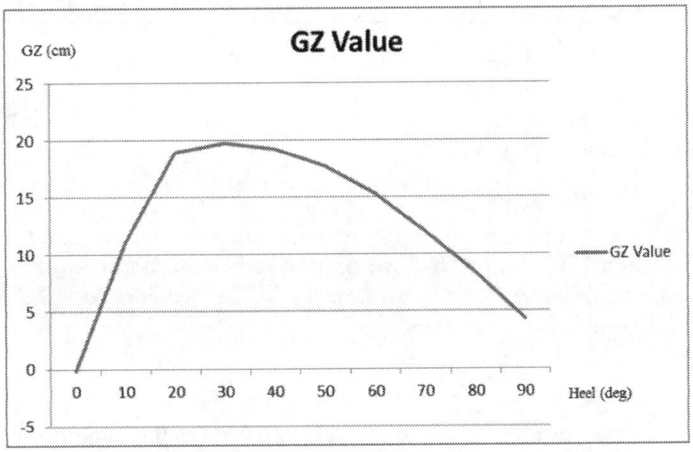

Fig. 2. The GZ curve of the SWATH Autonomous Surface Vehicle Hull Form

The curve as shown in **Fig. 2** is explained that the GZ value of ASV still has significant righting moment until 50° of the heel.

3.2 Intact stability characteristics

To predict the engine power of ASV, we must be calculating the resistance. For cruise speed (Vs) we assumed to 1 m/s or 1.94 knots. The resistance can be calculated with CFD method. It is necessary to calculate the drag coefficient of ASV. The drag coefficient is a dimensionless quantity that is used to quantify the drag or resistance of an object in the fluid environment. The drag coefficient is always associated with a particular surface area, [14]. The drag coefficient C_d is expressed as,

$$c_d = \frac{2F_d}{\rho u^2 A} \tag{1}$$

where, F_d is the drag force, which is by definition the force component in the direction of flow velocity. ρ is the mass density of the fluid. u is the flow speed of the object relative to the fluid. A is the reference area. The result is shown in **Table 2**,

Table 2. The total drag force at 1 m/s

Drag Force of ASV			
Reference Area (m2)	**Drag force of X-axis (N)**	**Drag force of Y-axis (N)**	**Drag force of Z-axis (N)**
8531.46	5.55444	0.284103	0.063112

Table 3. The result of Drag coefficient calculation

Drag Coefficient of ASV					
No.	**u (m/s)**	**u2**	**Fd (N)**	**A (m2)**	**Cd**
1	0.5	0.25	1.44998	0.853146	0.0133
2	1	1	5.55444	0.853146	0.0127
3	1.5	2.25	12.2469	0.853146	0.0124
				Average	0.013

As shown in **Table 3**, to obtain the drag coefficient, several simulations should be performed to ensure the estimated value. In this case, the three times simulation was performed with the variation of a service speed of the SWATH ASV. Furthermore, the obtained drag coefficient is about 0.013.

3.3 Seakeeping performance

Seakeeping ability is a measure of how well-suited a vessel is sailing the seaways environment conditions. To define that, we must predict the motions of ASV responses against its environment. The four principal terms needed to describe a seakeeping performance as follow, [15].

1. Mission: the role of the ship while at sea
2. Environment: the conditions under which the ship is operating (sea state)
3. Ship Responses: the response of the ship to the environmental conditions.
4. Seakeeping performance criteria: the established limits for the ship's responses.

The result could be calculated as the RMS (root mean square) of those motions, see **Table 4**. The root means square is defined as the square root of the mean square (the arithmetic mean of the squares of a set of numbers). According to the area under the response, spectrum delivers the mean of the squared responses, as that under the wave spectrum represent the mean of the derivation squared from zero line, [16]. Hence, the RMS motion is equal to the square of the area under response spectrum.
The roll motion could be interference the bathymetry sensor, which the distance of a transmitted sound would increase from the ASV to the seabed. Thus, the correction of this interference could be calculated as,

$$Y = \text{arcsec}\,\phi\,.\,h \tag{2}$$

where y is the extended distance of sound from the ASV to the seabed, ϕ is the roll motion and h is the distance of normal axis. This calculation can be the correction constant of the bathymetry sensor. Where, from the deviation of extended distance and normal distance, obtained the correction value. Thus, this correction value would be change along the roll motion value.

Table 4. The motions characteristic of ASV

Motions	90°	180°
Heave motion (m)	1.682E-01	6.198E-01
Roll motion (deg)	1.325E-02	4.291E-04
Pitch motion (deg)	7.393E-02	7.685E-02
Heave velocity (m/s)	1.264E-02	4.401E-02
Roll velocity (rad/s)	1.058E-03	6.120E-05
Pitch velocity (rad/s)	3.571E-03	1.765E-02
Heave acceleration (m/s2)	5.033E-05	2.920E-04
Roll acceleration (rad/s/s)	1.616E-06	1.764E-07
Pitch acceleration (rad/s/s)	4.265E-05	6.004E-05

4 Conclusions

The autonomous surface vehicles hull form design using Small Waterplane Area Twin Hull was made. The performance of the proposed ASV hull form has been evaluated according to the characteristics of intact stability, resistance, and seakeeping. Based on the result of the numerical analysis, the stability of ASV is accepted by the requirement of bathymetry survey activity. The minimum allowed GZ value is 25.0, and the actual GZ value is 26.4. The advantage of SWATH hull form is having a small resistance. This will save the power of the engine. And we can install a smaller engine. For Reynold number is about 9.97×10^8, and the ITTC experiment for $Rn=10^9$ is about 0.015 for drag coefficient. The ASV has smaller drag coefficient in the amount of 0.013. The motion behaviors would be limited the ASV cruise area only in the coastal area or inland waterways. The echo sound deviation result which is caused by roll motion could be calibrated by determining the roll angle during the depth measurement.

5 Acknowledgments

The work is funded by Directorate of Research and Community Services (DRPM) of the Indonesian Ministry of Research, Technology and Higher Education under the scheme of the excellent applied research of university - (Penelitian Terapan Unggulan Perguruan Tinggi) PTUPT program.

References

1. Iqbal M, Rindo G, KAPAL 12 19-24, (2015)
2. Schellenberger G., SWATH Technology Advanced SWATH Design Methods (ThyssenKrupp Marine Systems) (2007)
3. Baitis A E, Meyers W G, Woolaver D A, Lee C M, A seakeeping comparison between three monohulls, Two SWATH, and a Column-Stabilized Catamaran Design for The Same Mission (Naval Ship Research and Development Center), (1975)
4. Vaneck T, Manley J, Rodriguez C, Schmidt M, Journal of the Institute of Navigation 43, (1997)
5. U.S. Navy, Available online:www.navy.mil/navydata/technology/usvmppr.pdf (2007)
6. Curcio J, Leonard J, Vaganay J, Patrikalakis A, Bahr A, Battle D, Schmidt H, Grund M Proc. of Oceans 2005 (2005)

7. Ferreira H, Martins A, Dias A, Almeida C, Almeida J M, Silva E P, Encontro Científico - Robótica (2006)
8. Caccia M, Bono R, Bruzzone G, Bruzzone G, Spirandelli E, Veruggio G, Stortini A M, Capodaglio G, Robotics and Automation Magazine 12 95 (2005)
9. Ferreira H, Martins R, Marques E, Pinto J, Martins A O, Almeida J M, Sousa J B, Silva E P, Proc. of the Oceans Europe'07 Conference (2007)
10. Phillips S, Hook D, Young H, Proc. of UDT Europe 2008(2008)
11. Sager W, Shyu J, Manley J, Sea Technology 49 (2008)
12. Hine R, McGillivary P, Proc.of PACON 2007 (2007)
13. Benjamin M, Curcio J, Leonard J, Newman P, Journal of Field Robotics 23 (2006)
14. McCormick, Barnes W, Aerodynamics, Aeronautics, and Flight Mechanics, (John Wiley & Sons, Inc.) p 24 (1979)
15. Faltinsen O M, Hydrodynamics of High-Speed Marine Vehicles (Cambridge University Press) p 6 (2005)
16. Bhattacharyya R, Dynamics of Marine Vehicles (John Wiley & Sons. Inc, New York) (1978)

MATEC Web of Conferences **177**, 01002 (2018) https://doi.org/10.1051/matecconf/201817701002
ISOCEEN 2017

Development of Structural Dynamic Analysis Software of Truss Structure Using Monte Carlo Finite Element Method

Muhammad Ichsan[1,*], *Handayanu* [2] and *Agro* Wisudawan[2]

[1] Ocean Engineering Department of ITS, Undergraduate Student, 60111 Surabaya, Indonesia

[2]Ocean Engineering Department of ITS, Lecturer, 60111 Surabaya, Indonesia

Abstract. Structural reliability analysis is increasingly used in conjunction with the design of offshore structures which have high uncertainty. Two problems i.e. reliability and complexity of the structures resulted in Monte Carlo Finite Element Method (MCFEM) method of analysis. In the first part of this research, we will show results comparison between deterministic analysis software made in Proposed Software and analysis using commercial established software. In this case, object will assume as a simple Plane Truss structure with multi degrees of freedom applied dynamic load in the support of structure. The second part after obtaining a deterministic result compared by commercial software, further we will apply MCFEM algorithm to the deterministic analysis algorithm which has been made. The results of this research, contains a new procedure in the form of reliability simulation algorithm by combining Monte Carlo Simulation and Finite Element Method in Dynamic Model.

1 Introduction

Structural design should involve a various random variables both demand side such as environmental load, live load, etc. and supply side such as material strength, dimension, etc. In order to explain this random behaviour, statistics or probabilistic methods should be used [1].

In traditional structural analysis, structural strength and load will be simplified as the "maximum" (load) and the "minimum" (strength) which are largely based on deterministic data. This approach is not relevant since random behaviour of some structures are increase.

Some researchers have attempted to derive equations to predict these random variables in term of stochastic approach [2]. This approach will be a new reference to structural design procedure.

In the other hand, the complexity of structural geometry is not possible to be analyzed by using manual calculations. So, another helpful way to solve it, is by using computer-aided calculations which is known as numerical simulation [3].

The most widely used numerical method today is Finite Element Methods, but this method is limited to deterministic parameters used only [4]. Thus, to overcome this limitation, researchers try to combine finite element analysis and statistical methods to obtain efficient and effective results [5].

One of statistical method developed to solve random variables is Monte Carlo analysis [6] which is applied in deterministic analysis.

* Corresponding author: ichsan.celebes@gmail.com

On that background this research will discuss about Dynamic analysis on Simple Plane Truss structure by using Monte Carlo Finite Element Method Simulation.

In numerical computational analysis and design in science applications are known a lot of software such as Maple, Mathematica, Maxima, and so on. But at time went on Matlab came to dominate the market [7].

2 Research Description

In this study, the first attempt was to find relevant literature related to Monte Carlo, Finite Element Method on Dynamic of Structure, Proposed Software usage for finite element analysis, etc. Based on literature study, a dynamic displacement model calculation algorithm is created. Flowchart is a procedure that describes the steps of workflow in the form of flow diagrams. Flowchart calculations will be used to compile the function listings in Proposed Software.

After the displacement results in Proposed Software are known, then we will have performed the same analysis on the commercial established software to compare the results of previous calculations on the proposed software.

The creation of dynamic model analysis flowchart which is applied in Monte Carlo Finite Element Method then subsequently carried by forming Monte Carlo methods listing which will be added to the listing of displacement analysis of dynamic models that have been made previously.

After the analysis and determination are finished, it will be deduced from the problem and the previous goal to show how much structure reliability of the dynamic analysis are performed.

3 Result and Discussion

3.1 Reliability Analysis by using Monte Carlo Simulation and Finite Element Method on Dynamic Model Algorithm

Based on the results of literature study and review related to dynamic model structure analysis which will be applied *Monte Carlo* method, it is necessary some stage in determining the movement in the dynamic model structure of a simulation, such as determining the structure configuration based on the desired element type, perform the modal analysis and determine the method which is used in completing the dynamic model motion equation.

In the early stages, determine the distribution of random parameters which is used in the simulation process, as in Figure 1 random parameter consist of Young's modulus (E), Additional Mass (M_{add}), and Control Displacement (u_c).

After random variable is input, it must be repeated as much as the predetermined iteration by calling to 3 random parameters previously. Then, these 3 parameters are transformed to new variables E, M_{add}, and u_c which are used as inputs in calculations of simple Plane Truss Structures.

Dynamically model which is analyzed has assumption that acceleration happening in the support of the structure. Since real case known, a dynamic model analysis involves a fairly complex $p(\tau)$ function, the solution of the dynamic model equation uses the time history method with the completion of numerical techniques which is used the Newmark Method [8].

After the result is found as a displacement that occurs on each joint, it will be identified whether the element fails or not. The identification will be done by using Serviceability

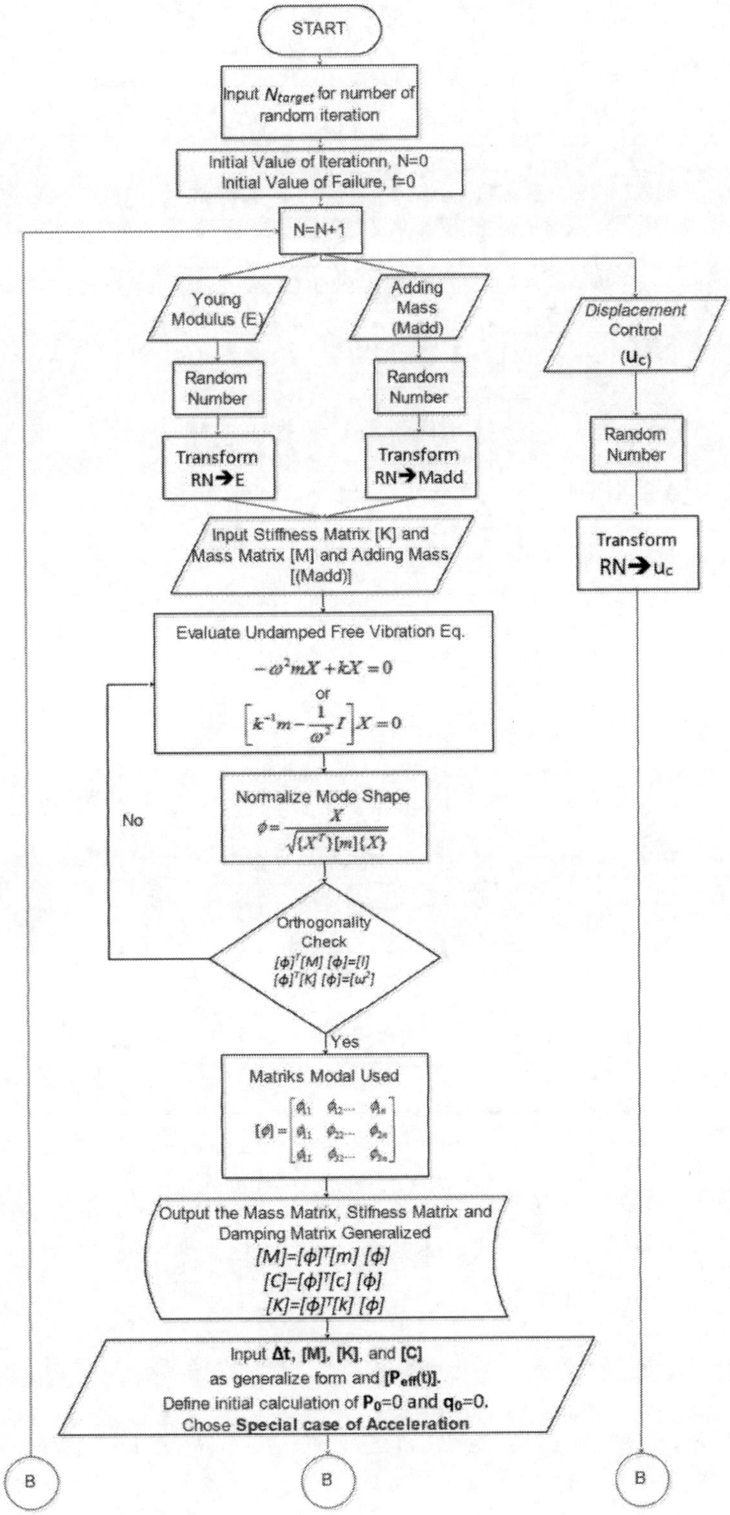

Fig. 1. Dynamic Model MCFEM Algorithm

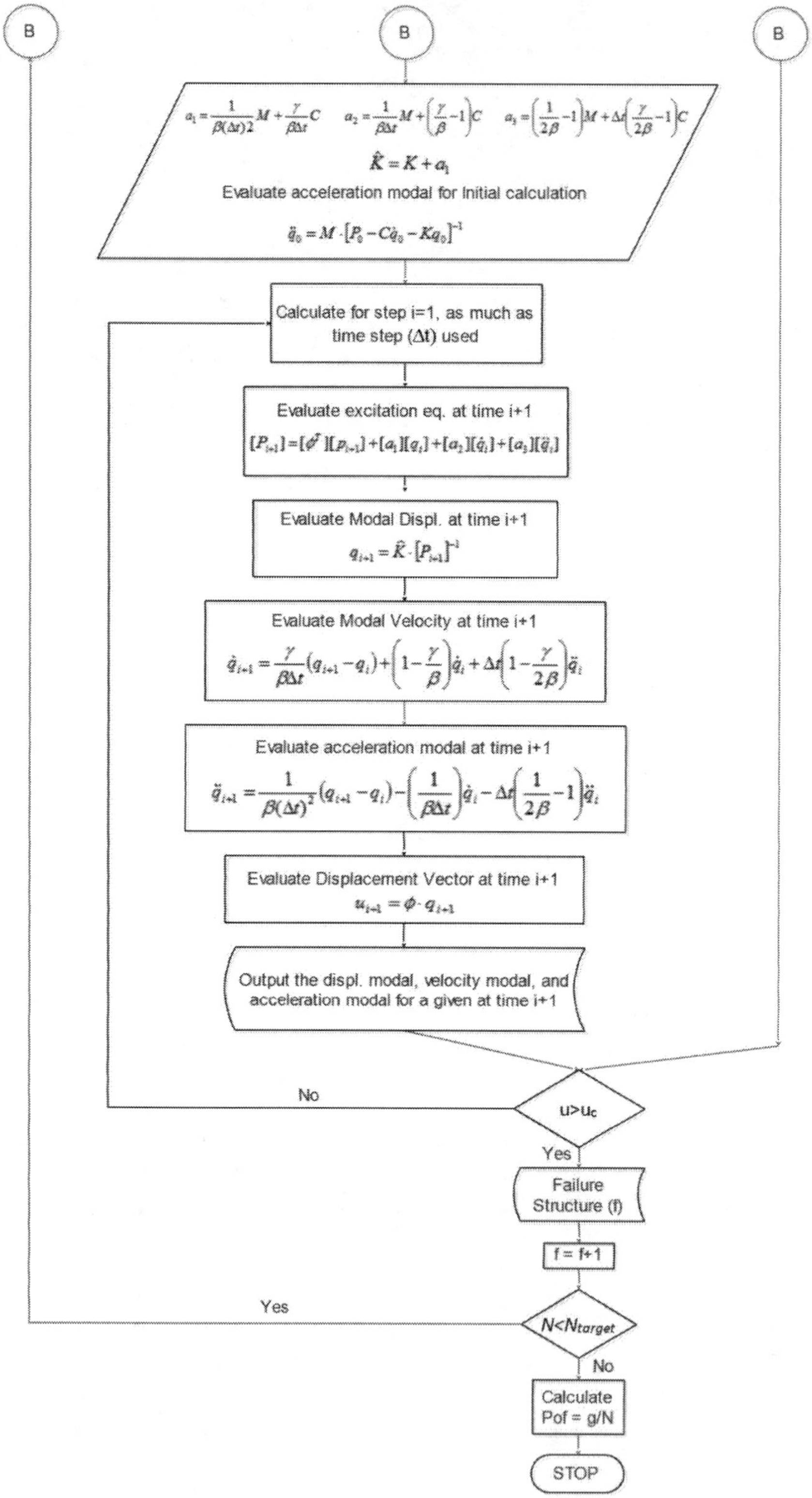

Fig. 1. Dynamic Model MCFEM Algorithm (Continued)

Limit State (SLS) method, the SLS provides the criterion that the structure keep standing with small earthquake intensity without getting damage [9] by using failure modes *MK(X)* as follows:

$$MK(X) \leq 0 \tag{1}$$

or

$$\Delta u_e \leq [\theta_e]h \text{ or } [\theta_e]h - \Delta u_e \geq 0 \tag{2}$$

or

$$u_{actual} - u_{control} \leq 0 \tag{3}$$

The equation 1-3 will be used for *probability of failure* calculation (P_f) as much as the predetermined simulation by using the equation below:

$$P_f = \frac{N_f}{N} \tag{4}$$

The number of failures (N_f) will be recorded for each $MK(X)<0$ and this simulation will be repeated as much as simulation (N) [10].
By P_f obtained, it can be determined the reliability of each *displacement joint* which is reviewed by using the equation below:

$$R = 1 - P_f \tag{5}$$

3.2 Monte Carlo App and Finite Element Method on Dynamic Model

There are several function listings module used in this research are:

a. Listing Function of Truss Element Stiffness
This function is used to determine the stiffness matrix of plane truss element on each element. The inputs to this function are Young's Modulus (E), cross sectional area (A), element length (L), and inclination angle.

b. Listing Function of the Truss Element Mass
In this study the mass matrix is assumed to be lumped parameter matrix (mass concentrated at each node). The input required in this function is the density, cross-sectional area and bar length.

c. Listing Function of Matrix Assembly
The assembly matrix listing is used to superimpose all elements of the stiffness matrix and the mass matrix of the structure into one stiffness matrix and one mass matrix only. The input of the above function is a zero matrix of size corresponding to the number of nodes, the stiffness matrix and the mass matrix of each element and the node which form elements (i and j).

d. Listing Function of Modal Analysis
Listing function is used to determine the natural frequency and shape modes of the model. The input of this listing is the mass matrix and the stiffness that is already in the form of matrix assembly and boundary conditions.

e. Acceleration type selection
Listing function is used to determine the type of acceleration desired based on the suitability of the data. The selection is between choosing "average" or "linear"

f. Listing Initial Condition
Listing function is used to determine the initial condition to be included in the time step process
This stage is divided into several parts, namely:
- Determine time step (Δt) used as based on the measurement data in the field of earthquake acceleration that occurred.

- Determination of initial conditions on values q0, , , U0, and P0.
- Determination , γ and β constants which follows the type of acceleration specified in the beginning.

g. Listing of time step calculation by using Newmark method.
h. Listing of Reliability Analysis
In this listing will be calculated the number of failures that occur for every simulation as much as the predefined simulation iteration.

3.3 Monte Carlo App and Finite Element Method on Dynamic Model

Software comparison done to find out how accurate the calculation of the simulation is, by comparing the proposed software with commercial software.
The model shown will be 2-dimensional simple Plane Truss by using lumped mass matrix. Figure Model 1 is as follows:

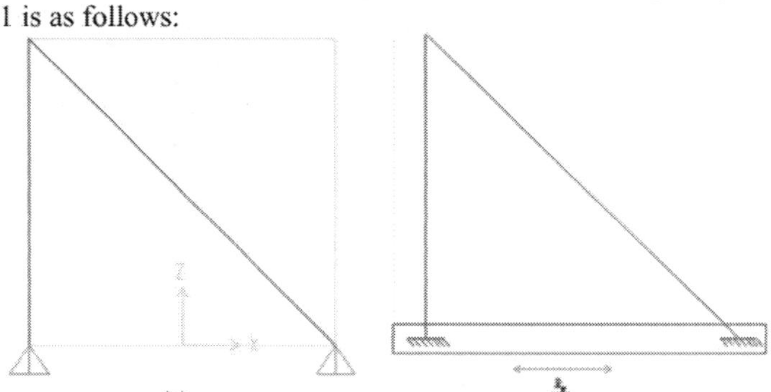

Fig. 2. Plane Truss Triangle Model

Representation of model shown above in the matrix is depicted in table 1 below:

Table 1. Relevant Parameters of Plane Truss Triangle Model

Mass Matrix, $\boldsymbol{M}$	$\boldsymbol{M} = (\mathbf{1.0e+05})\times\begin{bmatrix}1 & 0\\0 & 1\end{bmatrix}$
Matrix Damping, $\boldsymbol{C}$	$\boldsymbol{C} = \begin{bmatrix}0.6404 & 0\\0 & 0.6404\end{bmatrix}$
Stress matrix, $\boldsymbol{K}$	$\boldsymbol{K} = \begin{bmatrix}81.0270 & 0\\0 & 492.4916\end{bmatrix}$
Location of Influence Vector, $\boldsymbol{r}$	$\boldsymbol{r} = [1 \quad 0]$

The structure and acceleration properties of the analysis are as follows:

Table 2. Model Properties Triangle *Truss Plane* Material

Variables	Value
Young Modulus (E)	29000000 lb/in^2
Diameter (D)	12.3 in
Horizontal Element Length (L1)	118.11 in
Diagonal Element Length (L2)	167.0328 in
Mass type (ρ)	0.7 lb/in^3
Additional Mass (M_{add})	100000 lb

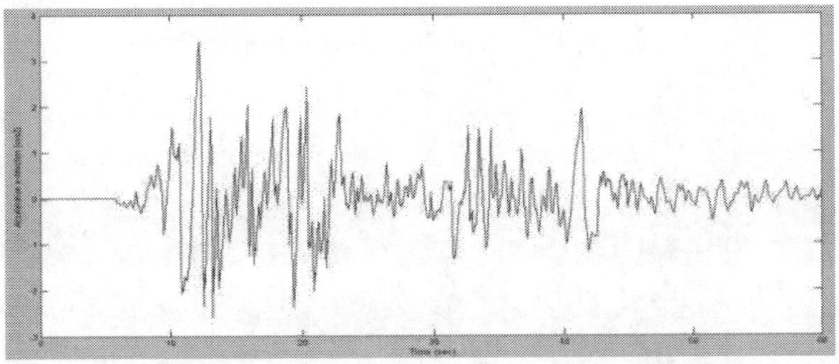

Fig. 3. Acceleration that occurs on the support of structure

The results of dynamic model analysis on the structure plane truss triangle model then compared each other. The comparisons are shown in table 3, table 4 and figure 4 below:

Table 3. Comparison of Natural Frequency Results between proposed software and commercial software

	Mode	Proposed Software	Commercial Software	Δ (Error)	Unit
Natural Frequency	1	9.001500	9.001500	0.000002%	rad/s
	2	22.192151	22.192000	0.000681%	

Table 4. Comparison of Maximum Displacement results between proposed software and commercial software

		Proposed Software		Commercial Software		Δ (Error)	
		(in)					
		Ux	Uz	Ux	Uz		
Dipl.	Max	9.86E-02	3.22E-02	9.01E-02	2.95E-02	8.532%	8.380%
	Min	-7.90E-02	-2.28E-02	-7.31E-02	-2.13E-02	7.453%	6.372%

Comparison of displacement graph shown in the picture below:

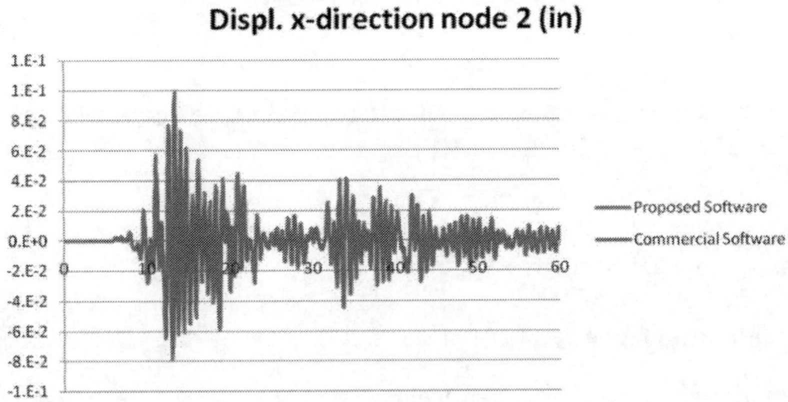

(a)

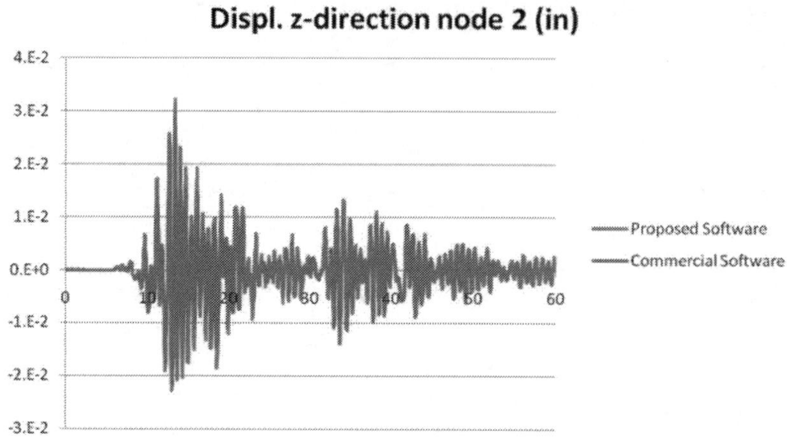

(b)

Fig. 4. Graph of Comparison Displacement node 2 (a) On the x-axis and (b) On the z-axis

The presented tables show that the differences between proposed software and commercial software are tolerable.

3.4 MCFEM application on Dynamic Model

In this steps, the reliability analysis by using Monte Carlo Finite Element Method is presented to Plane truss triangle model. The suggested simulation refers to Probability of Failure (P_f) [11] where the scale possible at the limit of failure does not occur at 10^{-5} otherwise according to [12] this simulation will be stopped if it meets the condition of convergence set (ε) that is shown in equations 6 and 7 below:

$$\frac{(u_{xi}+1)-u_{xi}}{u_{xi}}\times 100\% \leq 0.006\% \tag{6}$$

$$\frac{(R_i+1)-R_i}{R_i}\times 100\% \leq 0.6\% \tag{7}$$

Computer specifications used will affect the running time during simulation, while the simulation in this study will be done with a computer that has the following basic specifications:

- ✓ Processor = Intel® Core ™ i5 CPU @ 3.20 GHz
- ✓ RAM = 4.00 GB
- ✓ Operating System = Windows 8, 64-bit

3.5 Reliability Analysis Results

Plane Truss triangle model Structure in figure. 2 will be calculated by use the mechanical properties as follows:

Table 5. Mechanical Properties on Plane Truss Triangle Model Structure

Variable Input	Mean	COV	Distribution
Young Modulus (E)	29000000 lb/in^2	0.075	Normal
Yield Strength (σ_y)	39600 psi	0.078	Normal
Displacement Control (u_c)	0.161281241 in	0.078	Normal
Additional Mass (M_{add})	100000 lb	0.1	Normal
Diameter (D)	12.3 in	-	-
Horizontal Element Length (L1)	118.11 in	-	-
Diagonal Element Length (L2)	167.0328	-	-
Density (ρ)	0.7 lb/in^3	-	-

This calculation will be simulated gradually, so it will be known displacement and reliability of the system at each stage of the simulation. The simulation results can be seen in Table 6 below:

Table 6. Results of Running Structure of Plane Truss Triangle Model Structure

Number of Simulation	**Mean of Displacement**		**Reliability of System**	$\frac{(u_{yi}+1)-u_{yi}}{u_{yi}} \times 100\%$	$\frac{(R_i+1)-R_i}{R_i} \times 100\%$	**CPU Time**
	ux	**uz**				
10000	0.015833	0.004963	0.702000	0.000632	0.001	143.607
30000	0.015817	0.004958	0.854533	0.000632	0.001	426.201
80000	0.015822	0.004959	0.874613	0.000632	0.001	1091.061
100000	0.015818	0.004958	0.249420	0.000632	0.004	1452.644
120000	0.015815	0.004957	0.202317	0.000632	0.005	1748.010
150000	0.015829	0.004961	0.237560	0.000632	0.004	2181.845
180000	0.015826	0.004960	0.233411	0.000632	0.004	2543.478
200000	0.015825	0.004960	0.213390	0.000632	0.005	2809.365

From the table, it can be plotted the relation of the number of simulations with displacement and the reliability that occurs. As shown in figure 5 below:

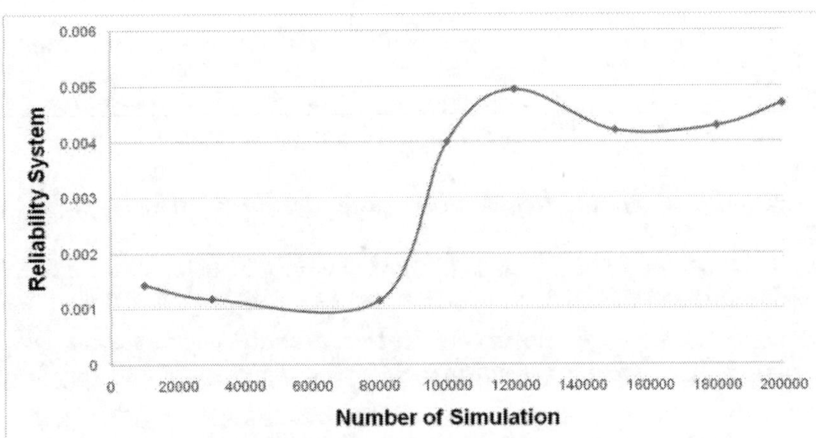

Fig. 5. Graph Relation between Simulation and Reliability

4 Conclusions

Based on the explanation of Monte Carlo Simulation application on Finite Element Method for Dynamic Model of 2 Dimensional Simple Structure above, it can be concluded as follows:

1. The proposed software has been compared with commercial software by using time history method with Newmark Method numerical technique. It showed the tolerable error.
2. Application of MCFEM as the solution of dynamic model equation has been made by using the MCFEM algorithm in Plane Truss Triangle Model. It is obtained the reliability 0.213390 in 200,000 times of simulations.

References

1. Schueller, GI, Shinozuka, M. ***Stochastic Methods in Structural Dynamics*** (Dordrecht: Netherland, Marthinus Nijhoff Publisher, 1987)
2. Cherng, Rwey-Hua, ***Stochastic Finite Element Analyzes of Uncertain Nonlinear Plane Under Random Excitatitions*** (Urbana: Illinois. National Science Foundation, 1992)
3. Chapra, SC, ***Numerical Method for Engineer. 6th ed.***(United States of America: McGraw-Hill, 2009)
4. Guoliang, J., Chen Lin & Dong Jiamei, RESS, 40, 77 – 83
5. Koduru, SD & Haukaas, T., Structural Safety, 32: 145-153, (2010)
6. Rosyid, DM, ***Pengantar Rekayasa Keandalan*** (Surabaya, Airlangga University Press, 2007)
7. Setiawan, A., ***Structural analysis,*** (Jakarta, Indonesia, Publisher Erlangga, 2015)
8. Campbell, SL, ***Modeling and Simulation in Matlab/Scicos with ScicosLab 4.4. 2nd ed,*** (London, Springer, 2010)
9. Chapra, SC, ***Numerical Method for Engineer. 6th ed*** (United States of America, McGraw-Hill, 2009)
10. Clough, Ray W & Joseph Penzen, ***Dynamics of Structures,*** (Berkeley, USA, Computers & Structures Inc, 2003)
11. Craig, Roy. ***Fundamental of Structural Dynamics*** (*New Jersey: Canada. John **Wiley,*** 2006)
12. Dymiotis, C., ***Reliability Based Calibration Code for Earthquake-Resistent Design,*** Edinburgh. JCSS Workshop on Reliability Based Code Calibration. (2001)
13. Ranganathan, R., ***Reliability Analysis and Design of Structures***(Delhi, Tata McGraw Hill, 1990)
14. DNV RP G101, Risk Based Inspection of Offshore Topsides Static Mechanical Equipment, (Norway: DET NORSKE VERITAS, 2002)
15. Graduates, A., Setyo, Y., Rosyid, D, ***International Conference on Technology, Informatics, Management, Engineering & Environment,*** Bandung: Indonesia, (2013)

The wave making resistance prediction of a mini-submarine by using tent function method

Aries Sulisetyono[1,*], and *Ardi* Nugroho Yulianto[2]

[1,2]Naval Architecture Department, Institut Teknologi Sepuluh Nopember, Indonesia

Abstract. This paper describes the wave making resistance solution of a mini submarine operating in under water surface with different level depth. The Thin ship theory was adopted to solve the problem for a case of the slenderness body. The source distribution along the centre plane of the body was expressed in Green's function of Havelock source potential under water surface. The Tent function method was proposed to illustrate the hull form based on offsets data, and to solve the Michell integral problem numerically. Four operational conditions were performed i.e. floating, snorkelling, and diving with 0.5m and 1m under water surface. The computational results for the mini submarine with length of 2m and diameter of 0.25m explained a more deeply operated under water surface cause to decrease a value of wave making resistance for all cases of Froude numbers. While in the diving conditions of 0.5m and 0.1m under the water surface, the wave making resistance were resulted about 64% and 74% less than the case of floating condition respectively. Furthermore, the effect of vertical fin on the body was investigated, where the wave making resistance could increase average 7.2% in snorkelling, 11.4% in 0.5m diving, and in the 1m diving about 9.07% for all Froude numbers. Over all the results of this approach shown a good agreement with the results come from Mitchell code.

1 Introduction

There were two total resistance components of ship such as a frictional resistance and a residuary resistance [1]. The frictional resistance of a hull ship was depend on the wetted surface area, Reynolds number, and form factor of hull. The wave-making resistance was the transfer of energy in the form of water waves, and manifests itself as a force opposing the forward motion.

The first theoretical solution of wave making resistance was given by [2] in term of the Integral equation. The Rankine source method was initiated to solve the Michell integral, in which the singularity have to be distributed on the water surface as well as the body surface [3]. Another method, the Havelock source needs only to be distributed on the body surface and it satisfied the free surface condition and far field conditions automatically [4]. The Michell integral equation was developed to calculate the wave making resistance on free surface conditions. For a submarine case, it was necessary to modify the Michell integral equation considers to a submarine mission under water surface [5].

In numerical computation, [6] introduce the 'tent' function to facilitate numerical integration over the ship hull for solving the wave resistance Michell integral. By using the tent function, the hull form can be easily expressed by the linear shape functions in terms of hull offsets that it is suitable for the case of single or multi hull ships. The wave-making resistance problem of multi-hull ships is solved by [7] using the tent function approach, and followed by [8] for single-hull fishing boat. The wave making resistance coefficient is used forward step in determining an optimized hull form of mini submarine case.

* Corresponding Author : sulisea@na.its.ac.id

The effect of vertical fin on the body need to be investigated in this paper since it device used to stabilize the platform of mini submarine. The stability and motion characteristics of a mini submarine was investigated by [9] numerically solving planar equations of a rigid-body motion subject to hydrodynamic force. The vertical fins effect due to the viscous resistance was predicted by using the wind tunnel test [10]. Generally, the appendage could increase the resistances value, but it had different effect consider to the operational condition of the mini submarine.

The intention of this paper was to propose the numerical approach of the tent functions in order to calculate the mave making resistance of under water vehicle. The effect of operasional depths under the free surface water was investigated as well as the the effect of vertical fin due to the wave making resistance.

2 Methodology

The body-fixed coordinate system for the problem was shown in Figure 1, Where *c* was steady advancing velocity, *D* was diameter, *Lwl* was length water line, *T* was draft, *Ss* was ship surface, and *Sf* was free surface. Assuming a fluid was an inviscid, incompressible, unbounded with free surface, and irrotational that satisfies Laplace's equation. Solutions satisfying particular boundary conditions could be superimposed to find the final solution based on the Green function method. The Green function which was the Havelock moving source have to satisfy the 1st order free surface boundary condition and the governing Laplace equation.

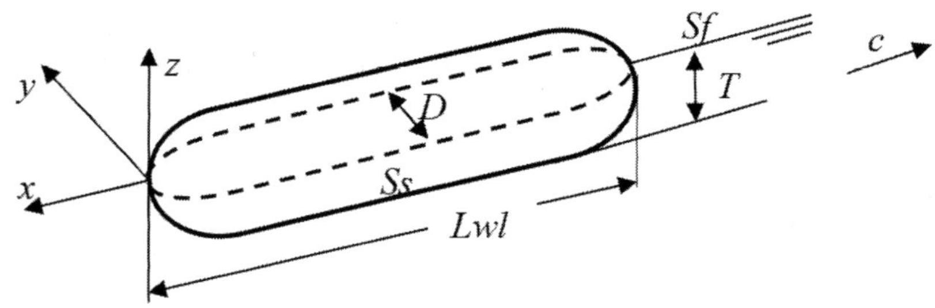

Fig. 1. Coordinate system

The wave resistance for a body was represented by a continuous distribution of sources and sinks moving in deep water was presented in equation 1 and 2 refer to [5].

$$R_w = 16\pi\rho k_o^2 \int_0^{\pi/2} \left[P(\theta)^2 + Q(\theta)^2 \right] \exp(-k_0 h \sec^2\theta) \sec^3\theta \, d\theta \tag{1}$$

$$\begin{Bmatrix} P(\theta) \\ Q(\theta) \end{Bmatrix} = \iint_S \sigma \ \exp(k_o z \sec^2\theta). \begin{Bmatrix} \sin \\ \cos \end{Bmatrix} (k_o x \sec\theta + k_o y \tan\theta \sec\theta) dS \tag{2}$$

Where $k_0 = g/c^2$, θ was the dummy variables for integration, and the source strength $\sigma = -cf_x(x,z)/2\pi$ was assumed over the ship hull center plane. For the case of a submarine that not only operates under free surface conditions but also in submerged conditions, an additional formula was required that indicates the effect of depth in operation with the addition of equations i.e. exp (- $k_0 h$ sec^2 θ). *h* was the depth of water, and the deeper the mini-submarine operates the smaller wave resistance due to the negative (-) sign.

For numerical solution, the tent function was used to facilitate the numerical integration as shown in equation 1 and equation 2. From the hull function *f*(*x,z*) was described under boundary condition, the hull slope function with respect to the x-direction could be represented as equation 3 and equation 4, and unit of tent function as shown in Figure 2.

$$f_x(x,z)=\sum_{ij} f_x^{ij}\frac{-c}{2\pi}(x,z) \tag{3}$$

$$\begin{aligned} f_x^{ij}(x,z)=&\left[\frac{1}{x_i-x_{i+1}}\right]\left[\frac{z-z_{j+1}}{z_j-z_{j+1}}\right]y_{i,j}+\left[\frac{1}{x_i-x_{i+1}}\right]\left[\frac{z-z_j}{z_{j+1}-z_j}\right]y_{i,j+1}\\ &+\left[\frac{1}{x_{i+1}-x_i}\right]\left[\frac{z-z_{j+1}}{z_j-z_{j+1}}\right]y_{i+1,j}+\left[\frac{1}{x_{i+1}-x_i}\right]\left[\frac{z-z_j}{z_{j+1}-z_j}\right]y_{i+1,j+1} \end{aligned} \tag{4}$$

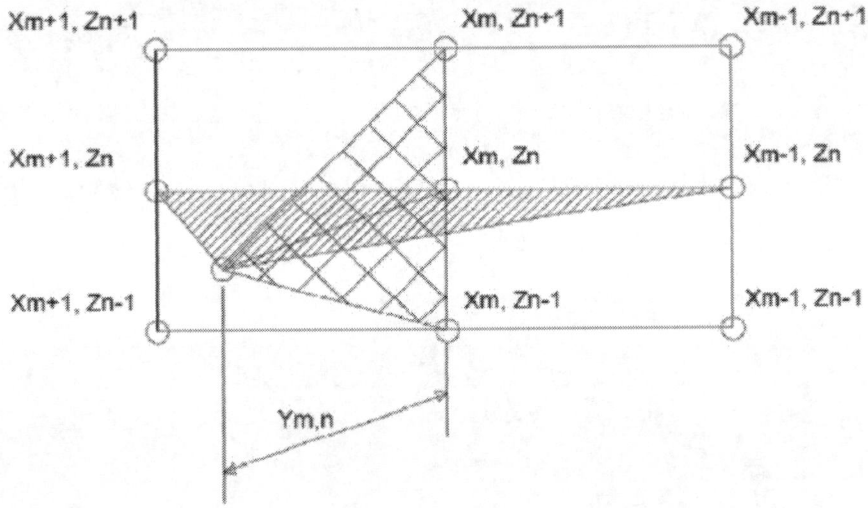

Fig. 2. Tent function unit

The P and Q in equation 2 could be expanded as,

$$\begin{aligned} \begin{Bmatrix} P(\theta) \\ Q(\theta) \end{Bmatrix} &= -\frac{c}{2\pi}\iint_S f_x(\mathrm{y,z})\ \exp\left(k_o z\sec^2\theta\right)\begin{Bmatrix}\sin\\ \cos\end{Bmatrix}(k_o x\sec\theta)ds \\ &= -\frac{c}{2\pi}\sum_{ij}\int_{X_i}^{X_{i+1}}\begin{Bmatrix}\sin\\ \cos\end{Bmatrix}(k_o x\sec\theta)dx \\ &\left\{A_{ij}\int_{Z_j}^{Z_{j+1}}\left(z-z_{j+1}\right)\exp\left(k_o z\sec^2\theta\right)dz+B_{ij}\int_{Z_j}^{Z_{j+1}}\left(z-z_j\right)\exp\left(k_o z\sec^2\theta\right)dz\right\} \end{aligned} \tag{5}$$

Letting the following equations represented each of the separate integrals gives, and hence

$$\begin{Bmatrix} P(\theta) \\ Q(\theta) \end{Bmatrix} = -\frac{c}{2\pi}\sum_{ij}\begin{Bmatrix}E_i\\ F_i\end{Bmatrix}\left(A_{ij}C_j+B_{ij}D_j\right) \tag{6}$$

Where

$$A_{ij} = \left[\frac{1}{x_i - x_{i+1}}\right]\left[\frac{1}{z_j - z_{j+1}}\right]y_{i,j} + \left[\frac{1}{x_{i+1} - x_i}\right]\left[\frac{1}{z_j - z_{j+1}}\right]y_{i+1,j} \tag{7}$$

$$B_{ij} = \left[\frac{1}{x_i - x_{i+1}}\right]\left[\frac{1}{z_{j+1} - z_j}\right]y_{i,j+1} + \left[\frac{1}{x_{i+1} - x_i}\right]\left[\frac{1}{z_{j+1} - z_j}\right]y_{i+1,j+1} \tag{8}$$

$$\begin{aligned} C_j &= \int_{Z_i}^{Z_{i+1}} \left(z - z_{j+1}\right)\exp\left(k_o z \sec\theta\right)dz \\ &= \frac{1}{k_o \sec^2\theta}\left\{\left(z_{j+1} - z_j\right) + \frac{1}{k_o \sec^2\theta}A_{ij}\right\}\exp\left(k_o z_j \sec\theta\right) - \frac{k_o z_j + \sec^2\theta}{k_o \sec^2\theta} \end{aligned} \tag{9}$$

$$\begin{aligned} D_j &= \int_{Z_i}^{Z_{i+1}} \left(z - z_j\right)\exp\left(k_o z \sec^2\theta\right)dz \\ &= \frac{1}{k_o \sec^2\theta}\left\{\left(z_{j+1} - z_j\right) + \frac{1}{k_o \sec^2\theta}\right\}\exp\left(k_o z_j \sec^2\theta\right) + \frac{\exp\left(k_o z_j \sec^2\theta\right)}{k_o \sec^2\theta} \end{aligned} \tag{10}$$

$$\begin{aligned} E_j &= \int_{X_i}^{X_{i+1}} \cos\left(k_o x \sec\theta\right)dx \\ &= \frac{1}{k_o \sec\theta}\left(\sin\left(k_o x_i + \sec^2\theta\right) - \sin\left(k_o x_i \sec\theta\right)\right) \end{aligned} \tag{11}$$

$$\begin{aligned} F_j &= \int_{X_i}^{X_{i+1}} \sin\left(k_o x \sec\theta\right)dx \\ &= \frac{1}{k_o \sec\theta}\left(\cos\left(k_o x_i \sec^2\theta\right) - \cos\left(k_o x_i + \sec\theta\right)\right) \end{aligned} \tag{12}$$

3 Numerical Results

The main body of a mini-submarine consisted of cylinder with length of body 2 m, and diameter 0.25, and the front body shape was a hemispherical. There were two vertical rudders at the behind of body to control body direction, and two side horizontal fins to increase the stability of body due to underwater current force. In this paper, the effect of vertical fin on the top of body was investigated, and two models were proposed including the mini-submarine with and without vertical fins as shown in Figure 3.

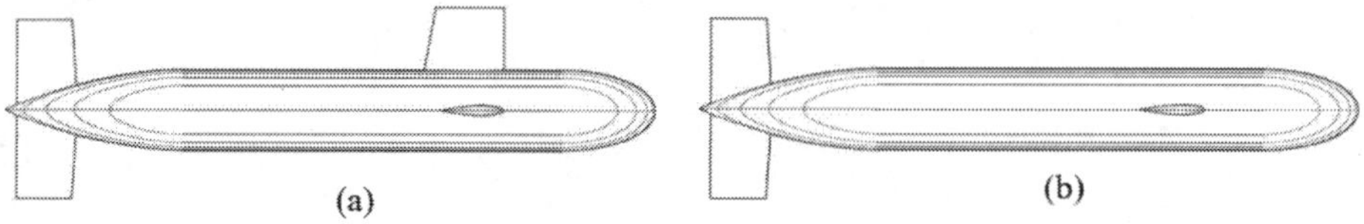

Fig. 3. The mini-submarine model (a) with, (b) without vertical fin

The computational were conducted in four cases of the operational conditions such as floating, snorkelling, diving 0.5m, and diving 1m. The five different speed were performed in terms of Froude number such as 0.23, 0.27, 0.34, 0.43, and 0.52 for all cases, and it results was compared to the Mitchell code which was developed by [5].

The computational results were shown in figure 4, where the numerical approach of Tent function (TF) produced generally a good agreement results with the Mitchell code (MI) for all cases of floating, snorkelling, diving 0.5m, and diving 1m on the configuration of a mini-submarine with vertical fin. Since the operational of a mini submarine was considered in low speed, the first hump of curve was occurred at Froude number 0.34, and the second hump would be occurred at highest Froude number. For the case of floating condition on the mini-submarine with vertical fin, the Tent function approach had results up to Mitchell code with average of 3.29% for all Froude Number. This differences value shown also for other cases of snorkelling, diving 0.5 m, and diving 1 m which had different value of wave making resistance coefficient about average of 4.45%, 10.01%, and 15.06% respectively.

In Figure 4, the mini-submarine without vertical fin had the results by using Tent function approach about average of 3.29%, 3.28%, 9.15%, and 14.66% higher than Mitchell code for all cases of floating, snorkelling, diving 0.5m, and diving 1m respectively.

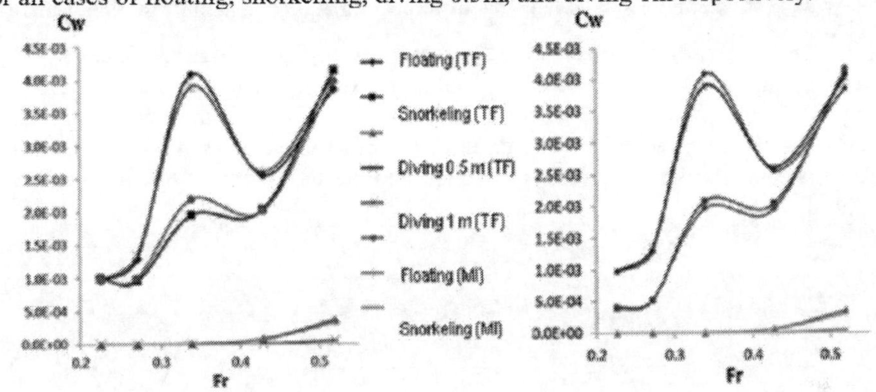

(a) (b)

Fig. 4. Wave making resistance results of Tent Function (TF) and Mitchell code (MI) for the mini-submarine (a) with and (b) without vertical fins

The wave patterns of four case conditions with Froude number 0.52 in which the diving condition of 0.5m and 1m had generate a small wave making as shown in Figure 5b. While the Froude number was decreased to 0.23, the elevation of wave surface was a very small for all cases of operational conditions as shown in Figure 5a. The value of wave making resistance might be zero if the mini-submarine operates at water depth more than 5 times of the ship's body diameter [11].

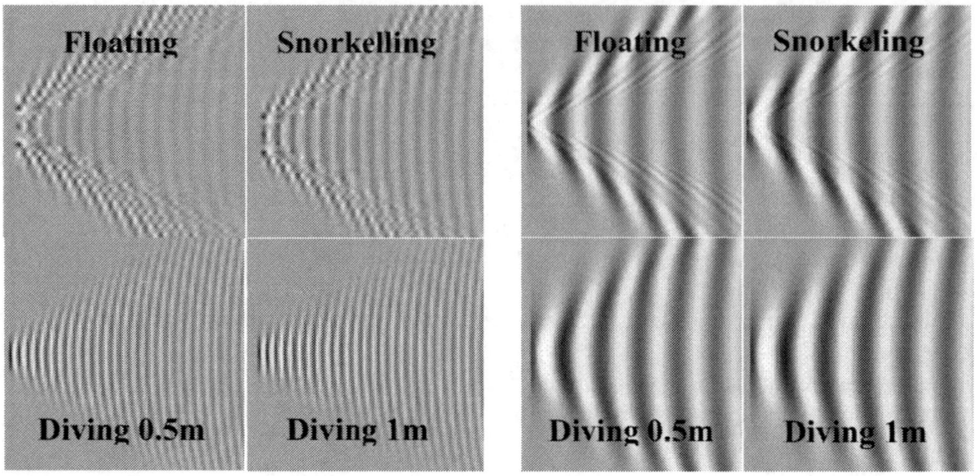

Fig. 5. Wave pattern of mini-submarine without vertical fin at Froude number (a) 0.23, and (b) 0.52

The effect of vertical fin on body was very small as shown in Figure 6. The wave making resistance coefficient of a mini-submarine with vertical fin was increased about average 7.2% in snorkelling condition, 1.4% in 0.5m diving condition, and in 1m diving condition about average 9.1% higher than a mini-submarine without vertical fin. The biggest different shown at moment of Froude number 0.23 while it operated in snorkelling condition.

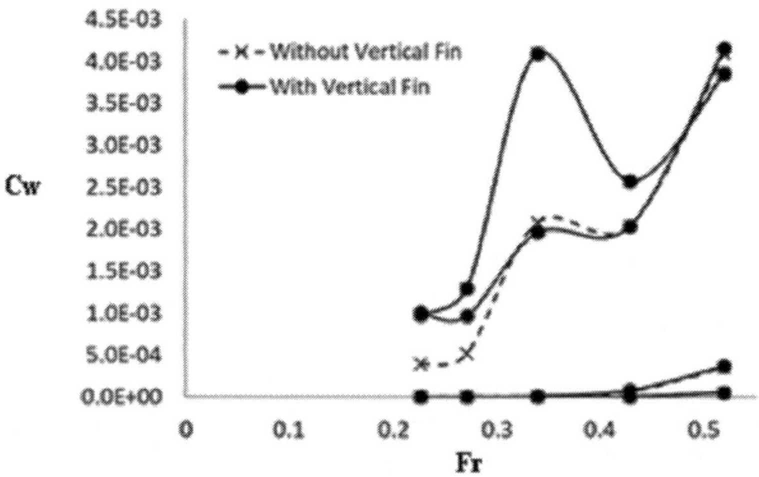

Fig. 6. Comparison results of Mini Submarine with and wihout vertical fin

4 Conclusions

In the diving conditions of 0.5m and 0.1m, the wave making resistance results were respectively about 64% and 74% less than the case of floating condition for all Froude numbers. The effect of vertical fin on the top of body was very small that could increase average 7.2% of the snorkelling condition, 11.4% of the 0.5m diving condition, and the 1m diving condition about 9.07%. Over all the results of the Ten Function approach shown a good agreement with the results came from Mitchell code.

References

1. W. Froude, The fundamental principles of the resistance of ships, *The Papers of William Froude The Institution of Naval architects* London (1955)
2. J. H. Michell, The wave-resistance of a ship, *Philosophical Magazine* Series 5 Vol **45** no 272 p 105-123 London (1898)
3. C. W. Dawson, A practical computer method for solving ship wave problems *Proceeding of the Second International Conference on Numerical Ship Hydrodynamics* p 30-38 University of California Berkeley (1977)
4. T. Havelock, The wave-making resistance of ships: theoretical and practical analysis, *The Collected Papers of Sir Thomas Havelock on Hydrodynamics Office,* Naval Research Department of the US Navy ONR/ACR **65** (1965)
5. E. O. Tuck, L. Lazauskas, and D. C. Scullen, Sea wave pattern evaluation (primary code and test results) University of Adelaide (1999)
6. C. C. Hsiung, Optimal ship forms for minimum wave resistance, *Journal of Ship Research* Vol **25** No 2 (1981)
7. H. Peng, Numerical computation of multi-hull ship resistance and motion, *Thesis at Dalhousie University* Halifax (2001)
8. A. Sulisetyono, Numerical computation for the wave making resistance prediction of ship without transom using tent function method, *Journal of Naval Architecture and Marine Engineering* (submitted) 2017
9. A. Sulisetyono and D. S. Purnomo, The mini-submarine design for monitoring of the pollutant and sewage discharge in coastal area, *Asian and Pacific Coasts 2009* p 252-258 Singapore (2009)
10. A. N. Yulianto, K. Suastika, A. Sulisetyono, Viscous-Resistance Calculation and Verification of Remotely Operated Inspection Submarine, *IPTEK The Journal for Technology and Science* Vol. **22** No. 4 (2014)
11. E.E. Allmendinger, Submersible Vehicles Systems Design, *Jersey: SNAME* (1990)

The Fortication of Iodine in Salt Consumption Using Manual Gun Spray

Intan Baroroh[1,*], *Bagiyo Suwasono*[2], *Didik Hardianto*[2]

[1]Department of Naval Architecture, Faculty of Engineering and Marine Science, Hang Tuah University, Indonesia
[2]Department of Naval Architecture, Faculty of Engineering and Marine Science, Hang Tuah University, Indonesia

Abstract. The Blue Economy concept begins with Artemia salt farming. The salt farming industry is one of the pillars supporting the economic of coastal community. However, in the market, most of salt consumed have iodine level that is less than 30 ppm. Therefore, it is necessary to measure the supply of iodized salt from the market to reach 30 - 80 ppm iodium according to Indonesia National Standard because it can impact our health. Thus, it was necessary to repair the iodized salt production equipment through an iodine spray device on the machine. The result of spray setting modification technology on macro ranged from 38 - 45 ppm which fulfilled Indonesia National Standard.

1 Introduction

The *Blue Economy* concept becomes the new trending topic for the government to explore the economical potency in oceanography and fishery in Indonesia developed to enhance the role of private parties in pro-environment economics development through business and innovative investment, and basically, there are five basic purposes to achieve; *natural resources efficiency, zero waste, social inclusiveness, cyclic systems of production: endless generation to regeneration, balancing production and consumption*, and *Open-ended innovation and adaptation*. This economics perspective is suitable with the program of poverty reduction (pro-poor), pro-growth, pro-jobless, and pro-environment. Thus, those five basic development, started with the Artemia salt farming become the milestone of the coastal economic community support. The distribution of the iodized salt from the company to the market or society depends on the production competency and marketing effort in free-trade era. The isolated areas with traditional markets will not be reached by the suppliers of iodized salt because of its few access. Thus, traditional markets mostly are supplied with non-iodized salt that encouraged people to fake the content of iodine. The conducted surveys showed that many packages of salt are claimed to have iodine in it, but the content of KI0 is less than 30 ppm as the standard applied.

* Corresponding author : intan.baroroh@hangtuah.ac.id

Based on a research [1] showed that there were 6 out of 10 types of salt sold in the markets in Bogor just contained iodine that were less than 40 ppm [2]. In food processing, based on a research done [3], using the food processor, the volume of the iodine added was less than the volume of the salt, so the distribution of the iodine was not spread well. Meanwhile, at Sunan Drajat Research Centre, the iodized salt machine is still far from the standardized one that is less than 30 ppm. Thus, it is very crucial to invent a device or technology, efficient and accurate, that meets the needs: the supply of iodized salt with the range of 30-80 ppm. The modified iodized salt spray machine will enhance salt farmers' productivity in the number of salt produced and its quality: iodized salt with the range of 30-80 ppm.

Some impact of iodized salt are very crucial for family health such as triggering the brain growth, making thyroid glands and fetus growth healthy, and the lack in consuming will lead to swollen thyroid glands, mental retardation, defected born baby, children with low smartness, miscarriage, etc. The excessive consumption will lead to hyper-thyroids, a condition when thyroid glands are produced excessively, and it forms goiter. In fact, goiter is also formed when thyroid glands are inactive, too. Moreover, sour taste emerges on the mouth and the saliva is produced excessively, and iodine also irritates the digestive systems by bruising it.

The usage of the gun spray on iodized salt device in the fulfilment of salt consumption for the salt farmers at Paciran, Lamongan will be illegible for the lab test, selling purposes, and people consumption, as the continuity of further purification process.

1.1 Working Principles of Manual Gun Spray to the Machine of Iodized Process for Soft Salt

The iodized device located at Sunan Drajat was still in defect stage since there was the iodine spraying process could not be done continually on the soft salt that results the changes of ppm, 25-50 ppm. Thus, it had to be repaired. Its modification was in the usage of manual gun spray using the energy of compressor for the first treatment done.

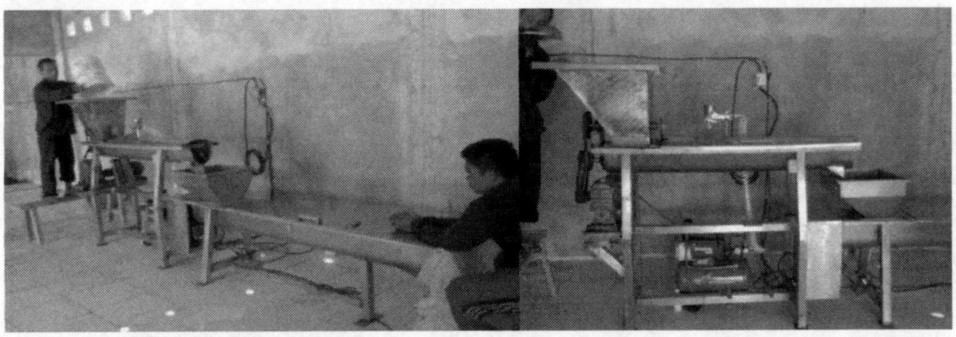

Fig. 1. The Existence of Iodized Machine.

The compressor was linked to the manual gun spray to fill the air from the compressor to the spray to make iodized clouds. The air from the compressor entered the spray by pushing the handle, and it was done continually. The air hose was so small, 3 mm in diameter, so it was easy to block because it was used by pushing it many times. At present, the iodized machine with its spray had already being attached and used at Raw Salt Processing Unit, Sunan Drajat, Lamongan to elevate the capacity and the quality of the product called Garam *Samudra*, shown on Picture 1: Iodized Machine**.** The mechanism was the soft salt with NaCl more than 97% content was put to the iodized machine regularly to filter the salt to the screw impeller bucket, and the machine was used to mix and spread iodine to the softened salt. Iodized salt product was produced using a machine with 2 types of parts: belt and screw conveyors within 2 stages. The first stage was putting the raw salt to the feeder, and then move upward through belt conveyor. At last, the final cycle was the iodine spraying process. At the second stage, iodized salt was moved from the belt conveyor to the screw conveyor that move upward to enter the container bucket. The weakness was stability and homogeneity of KIO3content in its each product. Iodized salt consumption must meet the standard, SNI 3556-2010 that has the content of NaCl>94% and the content of KIO3>30 ppm. Thus, the successful indicator of iodized soft salt fortification machine build-up was the achievement on the homogeneity level and the level of KIO3 that ranges between 30 to 80 ppm. The situation of iodized machine testing using brush water spray and compressor.

Fig. 2. The testing of brush water spray using a compressor. (left)
Fig. 3. The position of the brush water spray that was put into the screw of iodized machine (right)

2 Research Method

This research used a machine that applied two times of treatments on iodine by spraying iodized liquid through the nozzle spray to the device. The spray was intended to make iodized component (KIO3) that reached 30-80 ppm that becomes the standard of iodine consumption in Indonesia.

In the first measurement, it used spray nozzle with compressor power to spray iodine liquid to the smooth salt with the combination of 6 gram of iodine that was added by 120 cc of fresh water. Each trial needed 120 kg for five times, so the total weight was 600 kilograms.

In the second measurement, it used spray nozzle with compressor power to spray iodine liquid to the smooth salt with the combination of 6 gram of iodine that was added by 120 cc of fresh water. Each trial needed 100 kg for five times, so the total weight was 500 kilograms.

Both procedures above used the nozzle spray with a compressor power that was intended to make the salt smooth and homogeneous. Thus, the second procedure would make the result smoother than the first result and more homogeneous. At last, the resulted iodized salt would reach 30 - 80 ppm of consumptive salt.

Based on both final results, there were 30 salt samples that were tested using KIO3 at laboratory owned by Surabaya Airlangga University. Each step of the testing, it needed 120 kilograms of smooth salt that was applied within five times of stages, and 100 kg that also was applied within five times of stages. Thus, there were approximately 1,1 tons of smooth salt needed. If the process did not produce iodine component less than 30 ppm, the process was repeated from the rotary drying by mixing the smooth salt with iodized liquid concentrate based on its measurement until the sample showed the standardized iodine level. The flow chart of the research methodology can be seen at below.

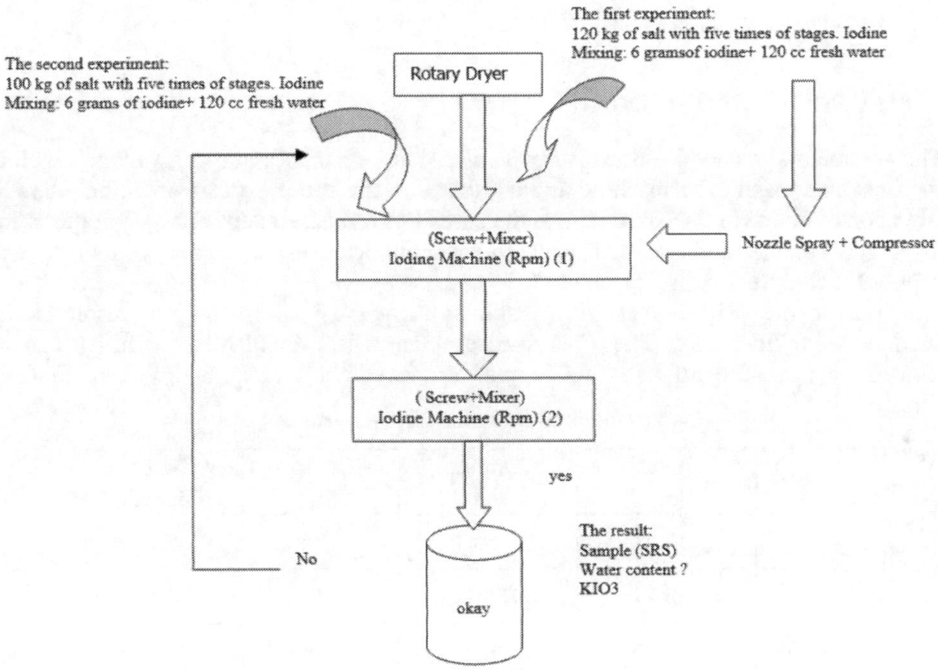

Fig. 4. The flow chart of research method of modified iodized spray machine.

2.1 The Indicator of Applied Research

The indicator of the research was 30 samples of KIO3 ranged from 30 – 80 ppm with water component less than 7%, and the adjusted compressor pressure on the device.
The expected final results were:

1. The whiteness of the salt was more than 4.7%
2. KIO3 component of the salt was 30 – 80 ppm
3. The content of the water was less than 7%.
4. KIO3 homogeny in each gram of produced salt.

2.2 The early study of iodized "*Garam Samudra*"

It needed a certain portion of iodized salt for the consumption that it ranged from 30-80 ppm that were spread well.

2.3 Iodized Spray Modification

Having the previous result, the modification was to measure the composition of iodine input and water to the salt to reach 30-80 ppm. The application of KIO3 (mg/kg) 30 - 80 ppm that was added by water <7%, using the balance of its pressure (Psi/bar)on the nozzle spray to the spinning of mixed salt (Rpm).

2.4 Chemical Lab Testing on Salt Consumption Product.

All samples taken were planned to have 32 samples using chemical lab testing to find out whether the salt final product met Indonesian Standard which is NaCl> 94,7%, KIO3 &

(mg/kg) 30–80 ppm, water content < 7% and its homogeny.

3 Result and Discussion

The research was conducted at Sunan Drajat, Lamongan, Banjarwati Village, Paciran Subdistrict, Lamongan. The machine applied there, in the process of fortification, showed unstable continuity of iodine spraying, so the spread of iodine content changes in the range of 25-50 ppm, so the repair on that machine was conducted using manual gun spray with the support of compressor energy.

The trial on the iodized machine capacity to reach the speed and the result of the production was undergone 5 times. The measurement of the input of the soft salt and its output was the decreased amount of the output salt that can be seen in the following table.

Table 1. Iodized Machine Capacity-testing.

No	Input (kg)	Output (kg)	The loss (kg)	Time (minutes)
1	112,88	106,64	6,24	5,4
2	114,55	116,32	-1,77	8,2
3	116,32	109,04	7,28	8,14
4	111,09	111,44	-0,35	8,15
5	111,44	110,59	0,85	8,08
Amount	566,28	554,03	12,25	37,97
Average	113,256	110,806	2,45	7,594

The result of the iodized salt machine was if the input was 113,256 kg, the output was 110, 806 kg, so it could be concluded that there was 2.45 kg of loss. The speed in the fortification process was 14.7 kg/minute, and its product capacity was 882 kg/hour.

For the trial of the device, manual gun spray and a compressor were used, and the spray of iodized liquid was applied by combining 6 grams of iodine that was added by 120 cc of plain water. Then, it was sprayed on soft salt of 120 kg for 5 times, so the total number of salt tested was 600 kg for the first trial. However, the second trial used the spray and compressor with 6 grams of iodine that was added by 120 cc of fresh water was the spraying to the soft salt, weighing 100 kilograms for five times, so the total amount of salt was 500 kilograms. Thus, the total number of soft salt used in this manual gun spray testing using compressor energy was 1100 kilograms.

3.1 The Testing Result of Manual Gun Spray Usage on Iodized Process of Soft Salt

The result of iodized–testing on the iodized machine was the sample result (conducted at Airlangga University lab) that has iodine content using spray and compressor as follows:

Table 2. Batch 1. Salt: 120 kg, iodine: 6 gr, and water: 120 cc

No	Batch 1	Iodine -1	Iodine -2	Iodine -3	Total	Mean	Deviation Std.
1	C1	12,47	10,12	9,41	32	10,66	1,60157
2	C2	60,84	60,29	61,91	183,04	61,01	0,82379
3	C3	74,12	74,19	75,61	223,92	74,64	0,84077
4	C4	17,88	14,77	14,64	47,29	15,76	1,83424
5	C5	32,86	31,44	29,73	94,03	31,34	1,56724
Total		198,17	190,81	191,30			
Mean		39,6340	38,1620	38,2600			
Deviation Standard		26,90718	28,13534	29,22898			

Table 3. Batch 2. Salt: 100 kg, iodine: 6 gr and water: 120 cc.

No	Batch 2	Iodine -1	Iodine -2	Iodine -3	Total	Mean	Deviation Std.
6	C6	85,59	85,27	86,21	257,07	85,69	0,47791
7	C7	42,39	42,25	42,08	126,72	42,24	0,15524
8	C8	52,46	52,54	52,66	157,66	52,55	0,10066
9	C9	23,23	22,88	22,46	68,57	22,85	0,38553
10	C10	23,67	20,84	20,42	64,93	21,64	1,76766
Total		227,34	223,78	223,83			
Mean		45,4680	44,7560	44,7660			
Deviation Standard		25,67985	26,26209	26,82080			

Production Target:
Indonesian Standard 3556 : 2010
Salt crystal > 94%
Iodine KIO3 > 30 ppm
Water (H2O) < 7%.

The Result of the National Standardized Salt was seen in the picture below.

Fig. 5. The Result of the Production of Samudra Salt at Islamic boarding school of Sunan Drajat, Lamongan

From the result of spray setting, it worked well on the second process; the preliminary and the final settings had a significant influence. The focus would be the pressure setting on the compressor with bar/Psi on the compressor. Moreover, nozzle setting was its height on the spinning of iodized machine screw that had to be reset because the hole was interrupted by its spinning in the iodine mixing process. Thus, the result of first testing result, Po (output) had to be put as the input, and the Pn had to be put as the input, especially on the material testing for C1, C4, C9, and C10.

4 Conclusion

In modifying the iodized spray device, we compared the water brush spray to several energy resources using a compressor. On tested-device stage 1, the test was conducted 5 times with the salt weight of 120 kilograms, and on the second stage, the salt weight was 100 kilograms with 5 times of test, so the total amount of salt used was ±1,1 tons. The preliminary and final settings showed that there was significant changes on the result, so it was crucial to set the pressure on the compressor. In addition, nozzle setting that showed the height of the nozzle on screw cycle on iodized machine needed to be reset because the holes of the nozzle was impacted by the iodized machine screw cycles when the iodine was mixed. The result of the trial as the first start salt machine was the Po (output) must be set back to the input, and the Pn (final result) must be set back.

The support comes from the Higher Education Board for the Implementation of Multi years of Reward Program and the decentralisation of Research Scheme on Applied Product of Higher Education for the batch of 2017 No. Ex.B/09/UHT.C7/V/2017 on May 4, 2017.

References

1. Komari et all, *The Journal of*Nutrition and food Research,(1995)
2. Pusriswilnon BRKP, The Guidance Book: *The Integrated Development of Salt and Artemia. The Research Centre of Sea and Resources. The Board of Oceanography and Fishery.* The Department of Oceanography and Fishery(2006).
3. Sulistyowati et al, "*Optimizing the Usage of Salt Fortification Result Mechanically and Chemically to have Iodized Salt*", Neptunus 19 no 1, ISSN. 0852-2812.(2013)
4. Utama. Rusda H, Oenzil F, Alioes Y , Journal of health andalas, **2** : 85-89 (2013).
5. Bagiyo Suwasono, Ali Munazid, Giman Hilmawan and Budtomo, *Design of Salt Fortification Mechanically and Chemically. Cooperation between Partners and KKP, Jakarta. UHT Surabaya and Oceanography Vocational School SunanDrajat, Lamongan.*(2010)
6. Amarullah, Husni and Sriyanto, *Technology on Artemia Salt and Other Related Products.* BPPT, Paper on a Workshop: The Future of Salt Industry in Indonesia. (2006).
7. Almatsier, Sunita, Basic Principle of Nutrition., Jakarta PT. Gramedia Pustaka (2003).
8. Ruwhof C and Drexhage HA, .*Iodine and thyroid autoimmun diseasein animal models.* Thyroid 11:427-36, (2001).
9. Anonymous, 1993, *Sodium Chloride in Chemieal Index.*
10. KerryMgruder, ,Halite, Guidelines forRoek Collection. (2000)
11. Bisnis Indonesia. *World Bank Gives Funds to 20 Salt Companies for SNI Processes,* (2000)
12. Bisnis Indonesia. *Salt Farming Needs Technology Standard,* (2000)

13. Fitriana, R. South Coast of Central Java Have Potential of Salt. Bisnis Indonesia, (2000)
14. Guidance Book of Iodized Salt. 2010. Salt Company, PT. Apel Merah , Rembang.
15. PT. Garam, *The Technology of the Making and the Problems of Salt Production.* The Department of Marine and Fishery.. (2000).
16. Rilleyand Skirrow, *Chemieal Oceanography*, Academie Press London, (1975)

Safety analysis of sea border transportation in indonesia, case study : countries between countries in sumatera (batam – singapore and batam – johor)

Danny Faturachman[1,*]

[1]Marine Engineering Department, Darma Persada University, Jl. Taman Malaka Selatan, Pondok Kelapa, Jakarta Timur, 13450, Indonesia

Abstract. As a border in Sumatera Island, Batam as a bridge between Indonesia and Singapore, and also Indonesia and Malaysia. Batam located in locations strategic considering in the sea border area, crossing between in Sumatera with Singapore and Johor, Malaysia. This research is aim to know whether the ASEAN Economic Community in order to ASEAN connectivity since 2015 application has already settle or not. .Because connectivity between ASEAN countries considered an important part in the context of creating the implementation of ASEAN economic community characterized by an open market in southeast Asia, therefore sea border transportation needs to develop again .This research is case study directly to see the fact in ferry routes between Batam – Singapore and also Batam – Johor and gain the information about the ferries, terminal and also the impact for the sea border transportation of three countries between Indonesia, Singapore and Malaysia.

1 Introduction

The development of transportation in Indonesia, especially sea transportation has been increased nowadays. This is the impact of economic activities and social experience culture and society. In addition, Indonesian sea waters besides as a link among cities and island, also among countries. As the border in Sumatera Island, Batam as a bridge between Indonesia, Singapore and Malaysia. Geographically, Riau Islands province with the boundaries of as follows; Northern: Vietnam and Cambodia, Southern: Bangka Belitung province and Jambi province, Western: Singapore, Malaysia and Riau province, Eastern: Malaysia and West Kalimantan province. In addition, Riau Islands province having smaller islands are almost spread across the city, including of them smaller islands that situated in the region of the borders of Indonesia.

The objective of the paper in order to safety analysis of sea border transportation crossing between countries especially between Indonesia, Singapore and Malaysia will look at how the crossing of the ship passing between that three countries. There will also be seen how safety standards which refers to SOLAS especially the safety equipment have been fulfilled on the ship. There will also be seen the condition of the crossing in port of Batam (in Sekupang, one of five ferry terminal in Batam), Harbor Front in Singapore and Stulang Laut, Johor Malaysia in terms of safety would have sufficient and appropriate for passengers. As a border in Sumatera island , the city of Batam as a bridge between Indonesia, Singapore and Malaysia .For all this necessary,

* Corresponding Author: fdanny30@yahoo.com

safety and security on cruise in the three countries for maritime safety affected business sustainable development especially in sea transportation activity. The method is case study with the data collection gained from the field on board between ferries on Batam – Singapore and Batam – Johor.

2 Literature Review

Indonesian's act no. 17 about cruise state that[6]:

- Ship safety and security is a state of the fulfillment safety requiresment and security in waters on transport, port and environment maritime;
- Ship safety is a state of vessel complient material, construction, ship engines and electricity, stability, general arrangement and equipment, a salvage supporter and radio, electronic of the ship, as evidenced by certificate post examination and testing.

2.1 Review Batam City

Batam City is one of municipal who are in the province of Riau Islands located at 0°25'9‖ – 1°15'00‖ northern latitudes and 103°34'35‖ – 104°26'4‖ east longitude. Batam City having broad territorial waters reached 1.570 km^2. The land area inhabited by 988.550 peoples, so that population density in the city as many as 38.661 persons / km^2. This population is the third populous of Sumatra Island after Medan City and Padang City. Batam City consisting of so many regions, there are Batu Ampar, Belakang Padang, Bulang, Galang, Lubuk Baja, Sei Beduk, Batu Aji, Segulung, Bengkong, the city of Batam and Sekupang [3]. Batam City is located very strategic island due to located along the international shipping with the limits of the area as follows: North Singapore and Malaysia, South Lingga district, West Karimun district, East the island of Bintan and Tanjung Pinang [1].

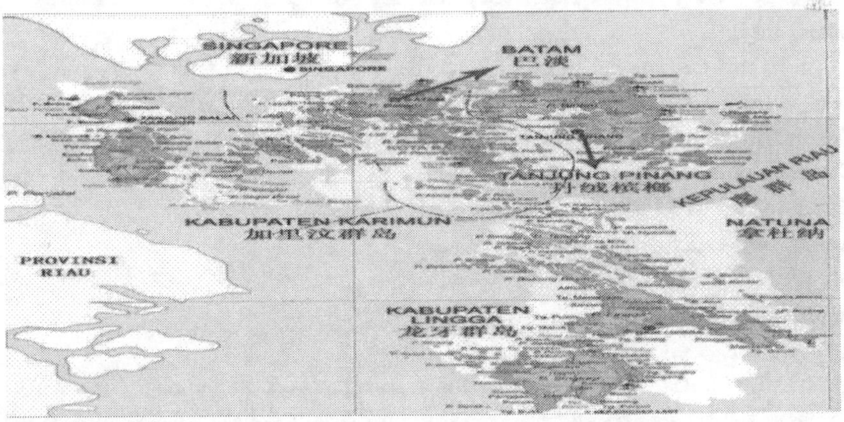

Fig. 1. Maps of Riau Islands
Source: http://www.indonesiatravelling.com

Batam City does not have natural resources which is abundant, hence economic activities the majority of the city depends on secondary sector and tertiary. This is reflected in the economic growth target the government of Batam that is driven by growth in the industrial sector and tourism. Batam who is regarded as the tropics, with an average temperature range from 24 up to 35 degrees centigrade (77 to 95 degrees Fahrenheit). Moisture in this area range from 73 % to 96 percent. In general the rainy season starts from November until April and the dry season from May to October. Mean annual rainfall about 2600 mm [4].

Transportation is a means of supporting mobility, where people can use the facilities in Batam public transport as taxi, buses, and motorcycles taxi. Besides transport, Batam that is also the islands regions, sea transport is one of the important transportation connected all the islands. The use of sea transportation connecting Batam with islands around and with neighboring countries as Singapore and Malaysia, make development and transport enough, as the ship ferry (ship crossing between countries).

2.2 Ferry Boat Classification

Ferry boat have some type and classified based on the way landing and also how loading and unloading a ship ferry [2]:

a. Based on the way landing .Landing way consisting of two kinds of:
   a) Ferry boat who land directly in the sand who named LCM (Landing Craft Manual) or LST (Landing Site Tanks). Access of the ship located at the upstream and the stern as well as having capacity conveyance greater.
   b) Another ferry boat that landed in a dock. A ship this requires a wharf for berthing. Access charge there are in the hull, upstream, and stern but access in the hull not common for their place on when docked in a dock. Access passengers being on the side, directly into deck side.

b. Loading and unloading based on the way. As a broad outline technology loading and fit on the ferry can vary as follows:
   a) Lo / Lo (Lift on / Lift off), namely vessel with the transfer of charge in vertical synchronization;
   b) Ro / Ro (Roll on / Roll off), namely vessel with the transfer of charge horizontally;
   c) Suction, namely types of vessels with the charge by means of suck / pump through a pipeline, usually combined with equipment conveyor belt (conveyor belt);
   d) Special, namely types of vessels who handles one kind of charge. To another ferry boat, because of the charge of people and vehicles and did not need equipment loading and unloading special, type suitable is Ro / Ro.

Included in the Ro/ Ro vessel type:

a) A short distance vessel

b) Intermediate distance vessel

c) Long distance vessel

The type of ship that sailed between Batam – Singapore and Batam – Johor are a short distance vessel called also fast ferry. This ship could called fast ferry because his speed in splitting the beach and the strait .Usually ships this type worn in the areas of waters or the sea no undulating high. So very suitable for transportation coastal streams and lakes that the waves not to strong. This type of ships much worn by a passenger ship that connects small islands. As is the case with Batam - Singapore, Batam - Malaysia, Batam - Tanjung Pinang, and Batam - Riau the mainland.

 https://doi.org/10.1051/matecconf/201817701005

Fig. 2. Example of Fast Ferry
Source: Own Collection Data

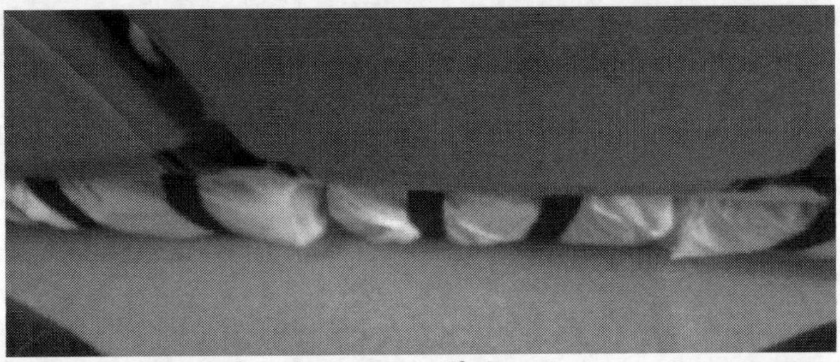

Fig. 3. Example of Life Jacket Below the Passenger Chair
Source: Own Collection Data

Fig. 4. A board Aids Discharging Life Jacket
Source: Own Collection Data

2.3 Ferry Terminal Classification

Consists of two words ferry and terminal, a point of terminal where passengers and goods in and out of the system and are essential components in the system of transport. While ferry means a vessel transportation close range [7]. Ferry terminal coordinates activities for accomodate passengers from dropping in and out of the ships at the close range or crossing boats. Also, terminal used for passenger activities from sub transit system to the sub differently characteristic of transit system. In other words means from sea transport to the land transport. Viewed from the scope of port system, ferry passenger terminal as a port component sub system serves for services inter island passenger ship. The ferry terminal components consisting of : 1. The pier, is a building used for tighted ferry for loading and unloading goods and passengers. The pier depends useful of port and the depth of cruise line, namely : elongated : a dock position parallel or parellel to the shoreline especially to cruise line that is deep enough for ship motion (ship maneuvering). 2. Wharf : a dock position overhanging to sea or perpendicular coastline. It was made when the depth of port waters lacking for the ship to go in and get maneuvering. 3. Pier : between dock and coastal connected with the bridge as being a sucessors for goods movement.

3 Result and Discussion

3.1 Sekupang Ferry Terminal, Batam, Indonesia

Sekupang ferry terminal is a national and international port, located on the west coast of Batam island. It connecting Batam with the island in the west side, Tanjung Balai Karimun (the Kundur island) and Dumai (the Bird island) as well as a group of island in the south like Bangka Belitung island, and also with the city of Palembang and Jakarta. For foreign country, Sekupang port have connencting transportation line with Singapore (Harbor Front) and Malaysia (Kukup). Sekupang located in the position of 01°10'24"east longitude and 104°00'06"south latitude. Port Sekupang in its function means of the departure and arrival for visitors, should be reflected the culture for the community surrounding, expected will could integrate with society as postive thing. In addition, Sekupang port is expected to provide complete services available to the community, example for the ticket sale of service, restaurant, the higher quality product and other facilities. These services expected to get the return back for Sekupang port from service in commercial of rental place for trade, advertisement, media and parking lot. Sekupang port have facilities like ferry terminal, parking area, building facilities, including special facilities for disabled.

Fig. 5. Ferry Terminal Sekupang, Batam
Source: Own Collection Data

References [5] regulated about the prot. The Sekupang port consisting of 2 floor : 1. The use of the steel structure in components column supporting areas, supporting glass area and the roof; 2. The separated of the circulation for the passengers (for arrival and departure in the first floor with the different of entrance; 3. Create a continuos space; 4 .Use structures wide lanscapes of space.

3.2 Harbor Front Ferry Terminal, Singapore

Harbor Front is the busiest port in Singapore, because here most pasengger from Batam, Tanjung Balai Karimun, Tanjung Pinang, down from cruise and entering Singapore. It is located in front of Sentosa island, where tourism came for. There are various entertainments ranging from tourist resorts, hotel, a golf course, underwater aquarium, including Universal Studios. Harbor front also have mal that sell various trinkets, starting from the clothes, electronic, food and others, so the Harbor Front is a fused port and are in the mall.

Fig. 6. Harbor Front Ferry Terminal Singapore
Source: Own Collection Data

3.3 Batam Centre Ferry Terminal, Indonesia

Batam Center Ferry Terminal. Batam Center is one of five international ferry terminal in Batam and the busiest terminal due to serve the route from Batam towards Singapore and Malaysia. Its location is in the center of town and the administrative center of Batam city, exactly in the way of Engku Puteri, Batam and adjacent to the hotel, the mall, and the center of Promotion of Sumatera, Batam Authority, the Mayor's Office, Immigration Office, Bank of Indonesia, Legislator Office, and other government offices in Batam. The interest of this terminal is the bridge connecting with the mall, namely Mega Mall, one of the largest malls in Batam, so tourist who come from Singapore or Malaysia can shopping there [8].

Fig. 7. The bridge in Batam Centre Ferry Terminal, Batam, Indonesia
Source: Own Collection Data

3.4 Port of Stulang Laut, Johor Bahru, Malaysia

Port of Stulang Laut, Johor, Malaysia is a modern harbor adjacent to the Mall is free of tax (duty free) so that tourists who want to shop for goods can buy anything with cheap price.

Fig. 8. Port of Stulang Laut, Johor Bahru, Malaysia
Source: Own Collection Data

There are two kinds of berthing off a ship which floats as well as four kinds of berthing others to support regular in the port of Stulang Laut. This port especially be able to accomodate until 8 ferries per hour.

Conclusion

1. For the ferries sailing from Batam to Singapore especially in Sekupang has ordered well and operating on schedule. Every port in Batam (5 ports) crossing does not cause density and mess the timetable. This because all ferries accomply with the deadline and crossing to Singapore also relative shortly only about 45 minutes once travelling.
   For fast ferries sailing the route Batam to Johor, Malaysia in particular Batam Center has arranged and operated on schedule. Although there are only 2 ports serving the departure to Malaysia, Batam Center and a new Harbor Bay, but the interest of tourists to pay a visit to and from Batam always increasing.
2. Safety analysis of the ship is already fullfillment and satisfy the requirements of safety according to the Indonesian regulation. Safety facilities in a basically enough available where life jacket there is enough and put under the passenger seat that can be used when the accident, so are the raft safety outside a ship can be directly used.
3. Sekupang ferry terminal has the international terminal facilities where it has separate the departure and arrival of passengers and also have a procedure security, where on the first floor is common zone, arrival of passengers and zona management with the different entrance. On the second floor is passengers departure zone.
   Batam Center ferry terminal have a decent facility as the international ferry terminal which has separated between departure and arrival passengers and also has good safety procedures as well as connected with a bridge straight to the mall, so that the tourist could be able to go shopping there.

The Harbor Front Singapore is amodern port have fused with malls and is a place of gate to entering the Sentosa island, which region of a tourist attraction and also busiest ferry terminal in Singapore.

Port of Stulang Laut, Johor, Malaysia is a modern port adjacent to the mall and free of tax (duty free) for tourist want to shop for goods can buy anything with the cheaper price.

References

1. Batam Industrial Development Authority, "Development Progress of Batam, First Semester of 2010", Batam, 2010.
2. Faturachman, Danny, "Analysis of Indonesian Sea Border Ferry Cross Countries, Case Study: The Comparison Between Indonesia and Malaysia.(Batam – Johor and Nunukan – Tawau), MASTIC, ITS Surabaya, 2018.
3. Faturachman, Danny, "Safety Analysis of Sea Transportation in Indonesia Case Study: Crossing Between Countries in Sumatera Island (Batam – Singapore), International Journal of Advances in Mechanical & AutomobileEngineering (IJAMAE) Vol. 3, Issue 1, 2016.
4. ICT Expo Step Forward to Batam Digital Island", Tribun Batam - Monday, 18th October 2010, accessed from http://batam.tribunnews.com
5. Indonesian Act No. 61 year of 2009 about Port.
6. Indonesian Legislation No. 17 year of 2008 about Shipping.
7. Morlok, Edward K, "Introduction Engineering and Transportation Planning", Erlangga Publisher, Jakarta, 1988.
8. Y.J. Naim, "Batam deserves to gain a high-growth economic incentives", Antara News Kepulauan Riau, 6th January 2011, taken from http://kepri.antaranews.com.

Stress Concentration Factor of A Two-Planar Double KT Tubular Joint due to In-Plane Bending Loading in Steel Offshore Structures

Rudi Walujo Prastianto[1,*], *Yoyok Setyo* Hadiwidodo[1] , and *Ibnu Fasyin* Fuadi[1]

[1]Department of Ocean Engineering, Institut Teknologi Sepuluh Nopember, Kampus ITS Sukolilo Surabaya 60111, Indonesia

Abstract. The purpose of this study is to investigate the proper Stress Concentration Factor (SCF) of a 60° two-planar DKT tubular joint of a tripod wellhead offshore structure. So far, calculation of SCF for a multi-plane tubular joint was based on the formulation for the simple/uniplanar tubular joints that yield in over/under prediction of the SCF of the joint. This situation in turn decreasing the accuracy of fatigue life prediction of the structures. The SCF is one of the most important parameters in the tubular joint fatigue analysis. The tubular joint is modelled as finite element models with bending loads acting on the braces that cover a wide range of dimensionless geometrical parameters (β, τ, γ). The effect of such parameters on the SCF distribution along the weld toe of braces and chord on the joint are investigated. Validation of the finite element model has shown good agreement to the global structural analysis results. The results of parametric studies show that the peak SCF mostly occurs at around crown 2 point of the outer central brace. The increase of the β leads to decrease the SCF. While the increase of the τ and γ leads to increase the SCF. The effect of parameter β and γ on the SCF are greater than the effect of parameter τ.

1 Introduction

Jacket platform is an offshore structure often used for oil and gas exploitation of offshore hydrocarbon reservoirs in shallow water areas, which have main structure consists of tubular members welded between braces and chords [1]. Stresses occurred at brace-to-chord intersection line of the tubular joint are very complex and unevenly distributed. The ratio between the maximum stress at the intersection line and the nominal stress in the brace is called Stress Concentration Factor (SCF). The SCF is one of the important components in the calculation of fatigue damage [2].

The majority of tubular joints commonly found in offshore structures are in the form of multi-planar tubular joints. The multi-planar effect plays an important role in the stress distribution at the brace-to-chord intersection areas of the tubular joints. For such multi-planar joints, the SCF formulae of simple uni-planar tubular joints are not applicable in

* Corresponding author: rudiwp@oe.its.ac.id

SCF prediction. The fact show that very few investigations have been reported due to the complexity and high cost involved, especially with bending loading conditions.

The significance of this study is to obtain proper SCF for a specific multi-planar double KT tubular joint. From design point of view this specific SCF would increase accuracy in fatigue life prediction for the joint itself. This research will study the effect of dimensionless geometrical parameters (β, τ, and γ) on the SCF distribution along the weld toe in the brace-to-chord intersection region of a multi-planar double KT tubular joint with several variations of the geometrical parameters subjected to in-plane loading at the braces. Finite element models of the tubular joint are used for local stress analysis to obtain stress distribution in the intersection region which is the accuracy obtained is strongly influenced by the numerical model to be analyzed [3].

2 Literatures Review

The understanding of the behaviour of completely multiplanar joints is very limited. During the past thirty years, a significant number of scientific papers and reports have been published on the study of SCFs in various uni-planar tubular joints. As a result, many parametric study in terms of the joint geometrical parameters have been proposed, providing SCF values of several uni-planar joint types for several loading conditions. Two following paragraphs briefly present the research papers on the determination of SCF distribution along the brace-to-chord intersection of uni-planar joints and on the calculation of SCFs in the multi-planar tubular joints, respectively.

Yeoh, et al. presented about behavior of tubular T-joints subjected to combined loadings [4]. Soh and Soh [5] researched about hot spot stress of K tubular joints subjected to combined loadings. Pang and Lee [6] presented three-dimensional finite element analysis of a tubular T-joint under combined axial and bending loading. Chang and Dover [7] predicted of stress distributions along intersection of tubular Y-joints and T-joints. Shao, et al. [8] predicted of hot spot stress distribution for tubular K-joint under basic loadings.

Research on the distribution of SCF on multiplanar tubular joints was done by Chiew, et al. [9], for the multiplanar tubular XX-joints. Karamos, et al. [10] derived the SCF formula and investegated stress distribution along intersection line on the multiplanar DT-joint due to the bending load. Research was also undertaken by Wingerde, et al. [11] formulating the SCF formula for uniplanar tubular joints K and multiplanar KK. Woghiren and Brennan [12] developed a set of parametric formulae to predict the values of SCF in multi-planar rack-stiffened tubular KK-joints. Naibaho, et al. [13] presented about SCF parametric study of multiplanar tubular KDT-joint. Puspaningtyas, et al. [14] researched about stress intensity factor parametric study of multiplanar tubular joint with fracture mechanics. Fuadi, et al. [15] derived the SCF formula for multiplanar tubular DKT-joint.

3 Finite Element Modelling

The structure to be analyzed in this case is a Multi-Planar Double KT tubular joint on the bracing of structure jacket platform (see Figure 1). Data used in this local analysis are structural geometry data, material properties and loading of the global analysis of structure jacket platform. The loading data is obtained from the global modeling of the structure (see Table 1 and Table 2). Chord and braces have the same material properties with the steel density and Young modulus of 563.5 lb/ft^3 and 30,000 ksi, respectively. Yield strength and Poisson's ratio of the steel are 50 ksi and 0.3, respectively.

In this multiplanar double KT tubular joint will be varied with several parameters, namely the parameters of beta (β), tau (τ), gamma (γ), and the angle between chord and brace (θ). The definition of each the tubular joint geometric parameters are: β (= d/D) is ratio between diameter of brace and chord; τ (= t/T) is ratio between wall thickness of brace and chord; γ (= D/2T) ratio between diameter and wall thickness of chord; and θ stands for angle between chord and brace.

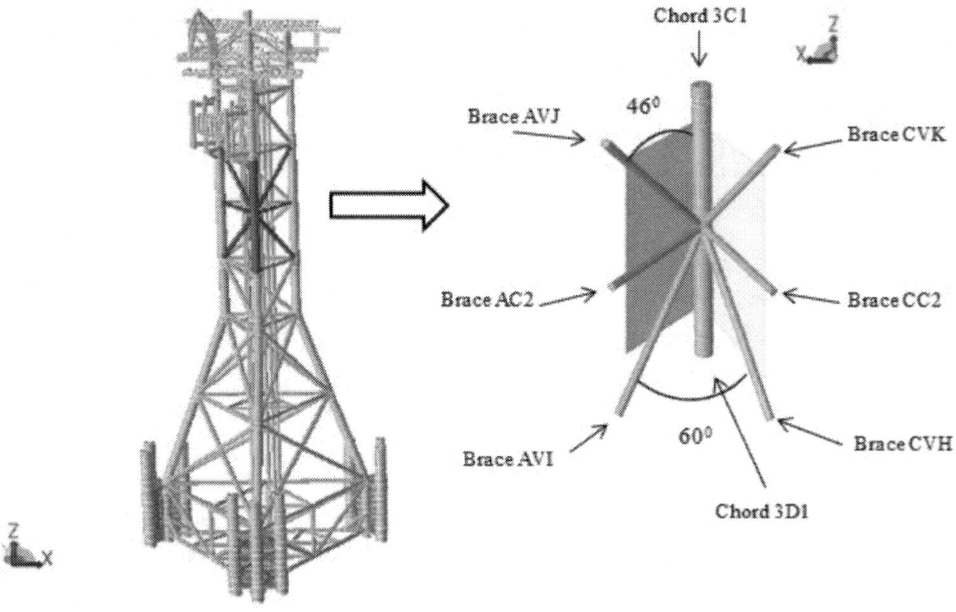

Fig. 1. A two-planar double KT (DKT) tubular joint taken from an offshore structure.

Table 1. Data of the DKT tubular joint.

Joint Member	Wall Thickness (in)	Outside Diameter (in)	Inside Diameter (in)	Length (in)
Chord 3C1	2	54	50	312
Chord 3D1	2	54	50	312
Brace CVK	1	24	22	540
Brace AVJ	1	24	22	540
Brace CC2	1	24	22	384
Brace AC2	1	24	22	384
Brace CVH	1	24	22	540
Brace AVI	1	24	22	540

As mentioned earlier, the aim of the present research is to study the effect of dimensionless geometrical parameters on the SCF distribution along the weld toe (on the chord side) of the main (inclined) brace. Different values assigned to each non-dimensional parameter are as follows: β = 0.35, 0.4, and 0.45; γ = 10, 12, and 15; τ = 0.35, 0.45, and 0.55. These values are selected to cover the practical range of the normalized parameters typically found in multi-planar tubular joints of offshore structures, namely: $0.3 \leq \beta \leq 0.5$; $0.3 \leq \tau \leq 0.6$; and $10 \leq \gamma \leq 14$.

Table 2. Moments and forces on the DTK tubular joint from global analysis

Joint Member	**Moments (kips-in)**			**Forces (kips)**		
	Mx	My	Mz	Fx	Fy	Fz
Brace CVK	-0.5	-32.9	-10	-48.6	-0.01	-0.1
Brace AVJ	2.3	-116.6	12.7	-13.5	-0.05	1.4
Brace CC2	213.4	-675.8	-5.4	9.9	0.01	9.8
Brace AC2	-94.6	-384.8	3.7	10.1	-0.01	5.6
Brace CVH	1.2	-42.9	20.5	49.7	0.04	-0.1
Brace AVI	-0.68	-186.3	31.4	-13.8	-0.11	1.6

Multi-planar double KT tubular joint to be analyzed is modeled by using a finite element method-based software with a geometry as depicted at Figure 2. Modeling is done by using the geometry, section and material properties and loading data as presented in Table 1 and Table 2. Weld modeling is necessary because the intersection between the two tubular (brace and chord) will form intersection line, where the stress concentration occurs in the area along the intersection line. The welding size along the brace-chord intersection satisfies the AWS specifications [16].

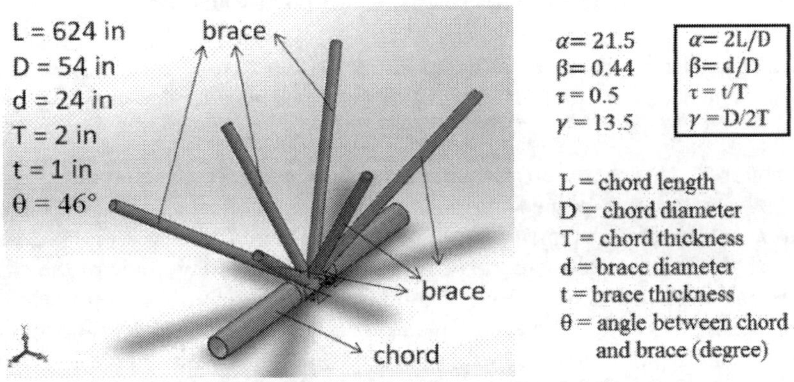

Fig. 2. Model geometry of the multi-planar DKT tubular joint for local analysis.

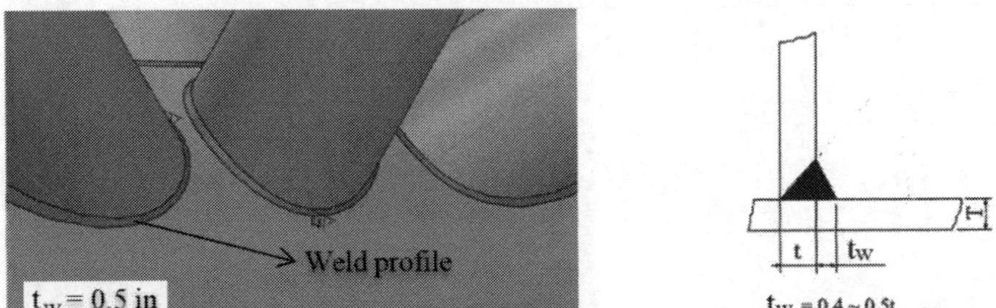

(a) Weld profile on the model (b) AWS D1.1. specifications for the weld

Fig. 3. Weld size along the brace-chord intersection satisfies the AWS specifications.

Meshing strategy for the tubular joint model is performed with meshing size is made smaller (smooth) on regions around the connection between the brace and chord. While in regions far from the connection area as observation areas, meshing size is made larger (coarser). The element used for the model is a solid three-dimensional element with a linear element type of tetrahedron (see Figure 4).

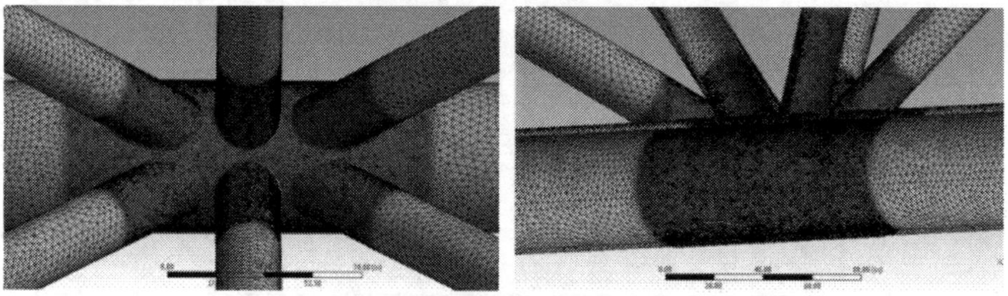

(a)

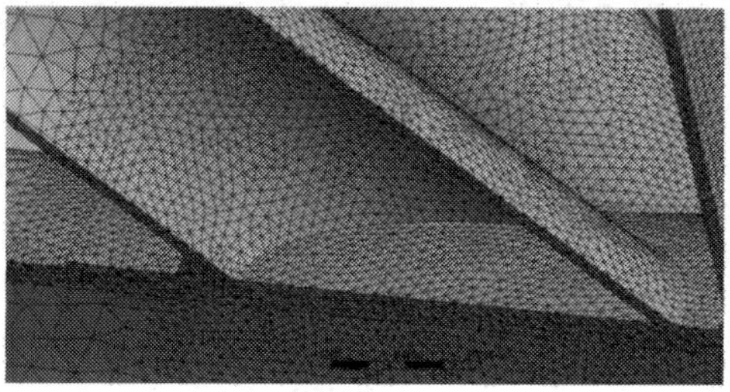

(b)

Fig. 4. Finite element model for the multiplanar DKT tubular joint: (a) Generated mesh for the tubular model, (b) Detail view of the generated mesh for the model.

The chord end fixity conditions of tubular joints in offshore structures may range from "almost fixed" to "almost pinned" with generally being closer to "almost fixed". For analysis in this study, it is assumed that all of the constraints are fixed on both ends of the chord. The tubular joint model with the parameter $\alpha > 12$, the magnitude of the stress at the joint area is not affected by the boundary conditions on the chord. Therefore In this study the value of parameters $\alpha = 21.5$ was assigned for all the models. The loading applied to the model are in-plane bending loading with the values obtained from the global modeling as presented in Table 2. In this study three different loading conditions (LC) are assigned to the models as can be seen in Figure 5.

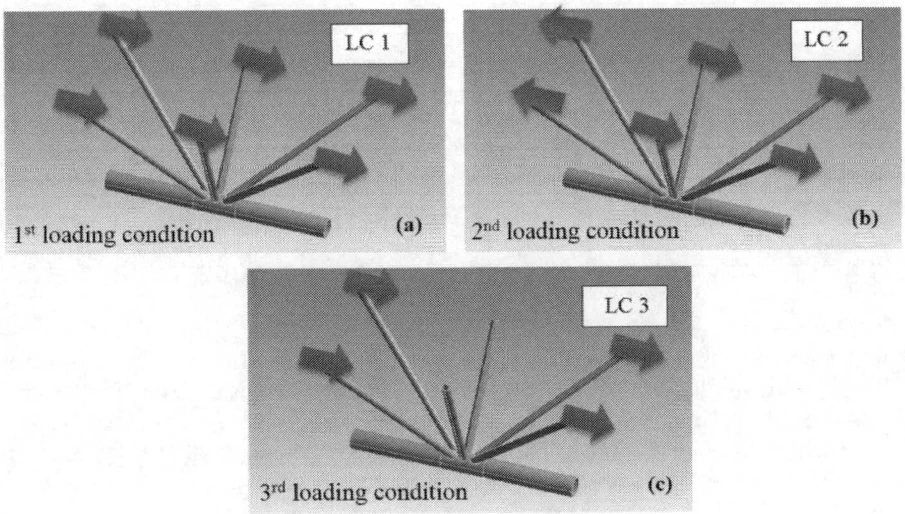

Fig. 5. Three in-plane bending (IPB) loading conditions for the model.

3.1 Model Validation

The meshing sensitivity analysis is performed to measure the consistency and accuracy of the results from the finite element model as a function of the number of elements used.

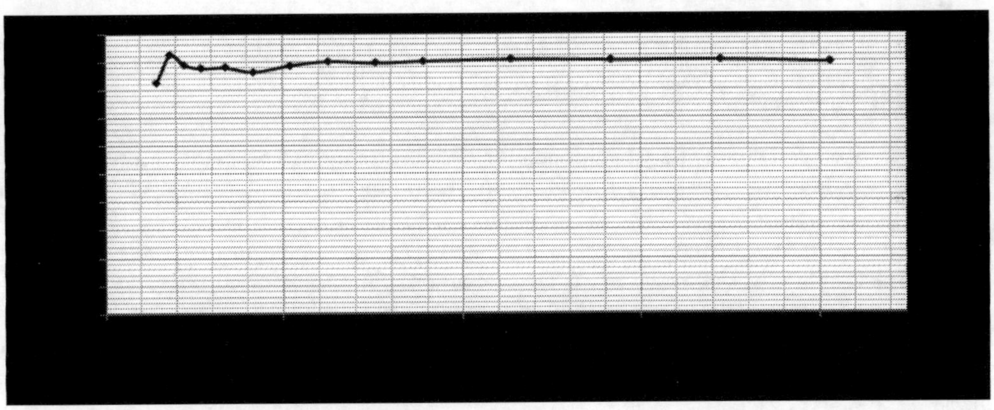

Fig. 6. Graph of meshing sensitivity analysis.

Size of meshing is changed iteratively, until the stress generated at a particular location becomes constant. From the results of meshing sensitivity test (see Figure 6), it is found that the number of meshing elements whose stress output is near constant at a particular point is meshing with the number of elements as many as 1,000,000 elements.

Table 3. Result of the meshing sensitivity test.

No	Number of elements	CPU time (minute)	Disk Storage (bytes)	Stress (ksi)	Change (%)
1	358,179	10	1,168,567,338	8.261	
2	373,059	10.2	1,179,997,513	9.235	10.547
3	390,022	10.3	1,195,417,876	8.877	4.037
4	408,545	10.5	1,205,657,858	8.763	1.301

5	435,671	10.8	1,217,451,712	8.788	0.284
6	467,794	11.3	1,220,329,118	8.633	1.795
7	509,145	11.7	1,239,971,901	8.859	2.551
8	551,325	12.1	1,245,735,142	8.992	1.479
9	603,822	12.6	1,250,661,018	8.933	0.660
10	656,583	13	1,260,421,134	8.991	0.645
11	754,722	13.6	1,277,445,914	9.058	0.740
12	867,228	14.1	1,282,657,781	9.025	0.366
13	990,615	14.7	1,295,957,708	9.029	0044
14	1,111,926	15.5	1,311,581,327	8.942	0.973

Another way to validate the results of the local analysis for the model is to compare it with the global model in term of the nominal stress in the brace using Von Mises stress, as presented at Table 4. From Table 4 it is shown that the difference for both analysis is very small (less than 1%), indicating that the model for local analysis was accurate in terms of proper structural modeling, loading, boundary conditions applied and also the results of the analysis.

Table 4. Nominal Stress validation.

$\sigma_{N.Von\ Mises}$ (ksi)		Error (%)
FEA (Global model)	FEA (Local model)	
9.0297	9.0298	0.000916

3.2 Hot-spot Stress and SCF Calculation

The widely accepted conventional approach for fatigue strength assessment of tubular joints is to use the geometric stresses at the weld toe. According to DNVGL-RP-C203 [17] hot-spot stress determined using linear extrapolation. The linear extrapolation method is done by making two points forming an area which will be extrapolated in linear stress. In this study, to determine the distance of two linear extrapolation points used refers to DNVGL-RP-C203 (see Figure 7). The maximum principal stress is used as linear extrapolation to determine the hot-spot stress [18].

After the known nominal stress and hot-spot stress, the formula stress concentration factor is:

$$SCF = \frac{\sigma_{maks}}{\sigma_n} \tag{1}$$

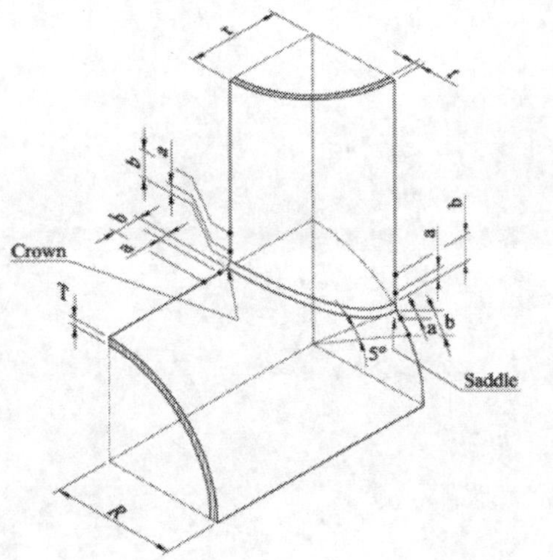

Fig. 7. Extrapolation procedure recommended by DNVGL-RP-C203.

4 Results and Discussion

The result of the parametric study with finite element approach is performed to determine the effect of the non-dimensional parameter variation on the stress distribution along the intersection line between the brace and the chord, which in present study is done along the weld toe on the chord side. Stress distribution along the weld toe on the chord side to be investigated only on the joint between the brace and the chord that experienced the greatest stress.

In this study, a brace of the DKT that experienced the greatest stress due to the in-plane bending loading is one of the central braces (the CC2 brace) as can be seen in Figures 8 and 9. It is very important to obtain stress distribution (then presented as SCF distribution) along the brace-chord intersection line of the observed brace. It was used 8 points to describe SCF distribution along the intersection line (see Figure 10) with interval of 45°. Point 1 (as Crown 1 point) is started at a position of an angle of 0°, the second and other next points are located at every next increment of 45° in clockwise direction. The Point 3, 5 and 7 is then called as Outer Saddle point (at 90°), Crown 2 point (at 180°), and Inner Saddle (at 270°), respectively.

To evaluate effects of the dimensionless geometrical parameters on the SCF distribution at the weld toe (on the chord side) along the intersection line of the main (central) brace, different values are assigned to each non-dimensional parameter, namely $\beta = 0.35$, $\beta = 0.4$, and $\beta = 0.45$; $\gamma = 10$, $\gamma = 12$, and $\gamma = 15$; and $\tau = 0.35$, $\tau = 0.45$, and $\tau = 0.55$.

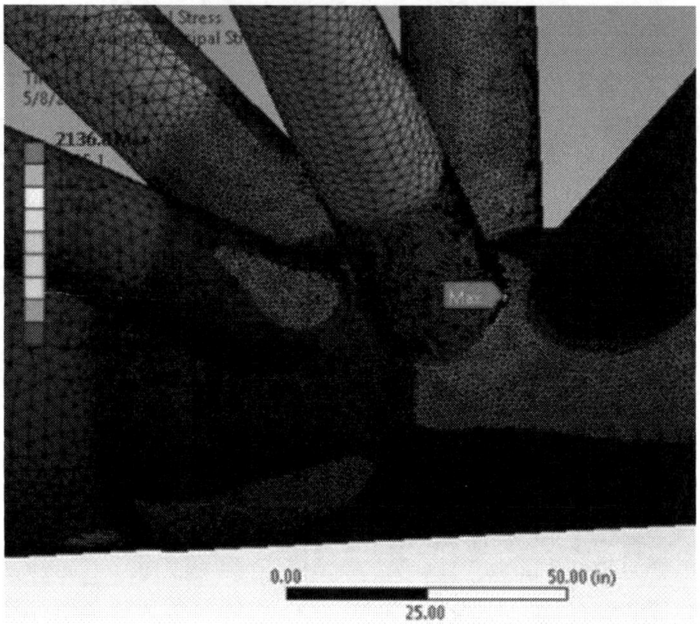

Fig. 8. Maximum stress occurs at the central brace-CC2.

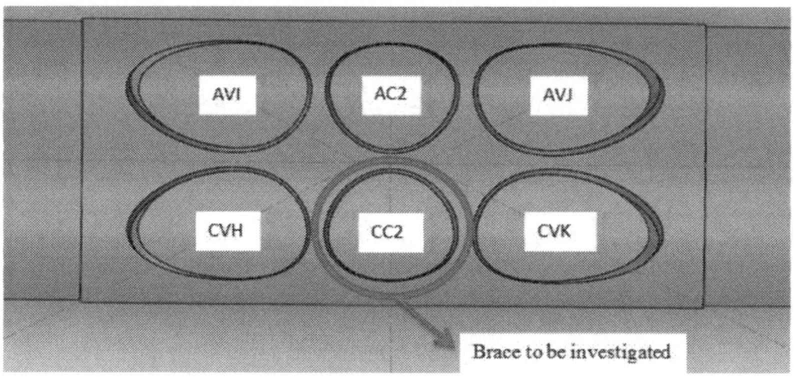

Fig. 9. Location of the central brace-CC2 to be investigated.

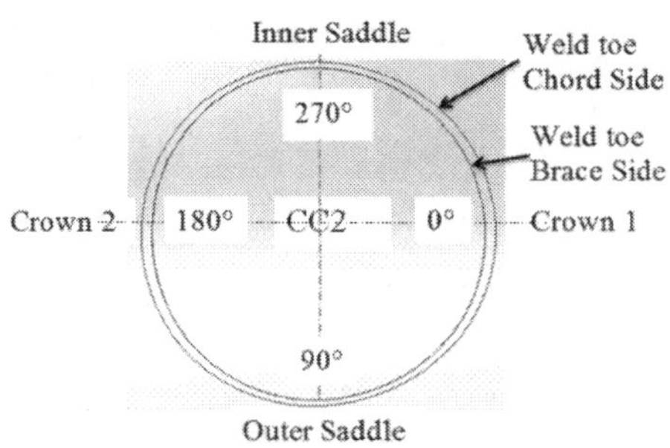

Fig. 10. Observed positions in the brace-chord intersection line of the central brace of the model.

4.1 Effect of β on the SCF Distribution

The parameter β is a ratio of the brace to chord diameter. In this study, there are three models with different of β for each model, i.e. $\beta = 0.35$, $\beta = 0.4$ and $\beta = 0.45$, respectively. Increase of the β in the models having constant value of chord diameter leads to increase the brace diameter. All parameters except the β parameter, i.e., the parameters τ, γ and θ are made equal for the three models ($\tau = 0.5$, $\gamma = 13.5$, and $\theta = 46°$).

Figure 11 shows results of SCF distribution due to three different load conditions. For all applied in-plane bending (IPB) load conditions, the peak SCF majority occurs in the area around the point of crown 2. For the case of IPB-LC 1 the peak SCF occurs almost precisely in crown 2 point for all β values. But two SCF peaks appeared at positions of around 135° and 225° for $\beta = 0.35$ at the case of IPB-LC 2. Very different pattern of SCF distribution along the brace-chord intersection line occurred for the case of IPB-LC 3 rather than the two previous loading conditions. For the latter case, two peaks of SCF appeared which is the first peak occurred at 135° and the second at 225° for all three β values. For smaller β, the difference between two peaks of the SCF becomes much larger.

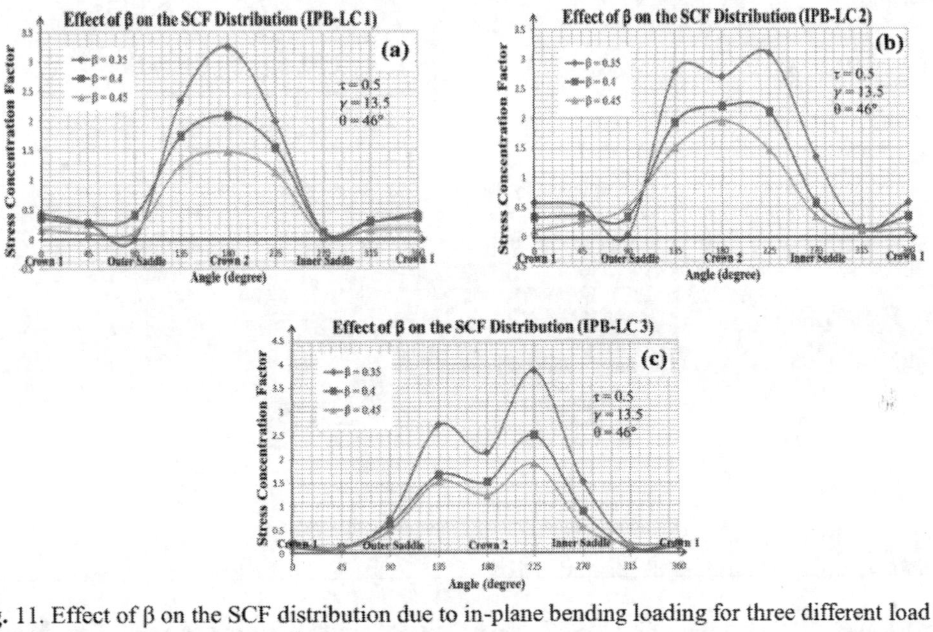

Fig. 11. Effect of β on the SCF distribution due to in-plane bending loading for three different load conditions (LC 1, LC 2, and LC 3).

Generally, results of the investigation indicate that increase of the β leads to decrease SCF around the Crown 2 area along the weld toe (on the chord side) at the brace-chord intersection line. The largest SCF of about 4 occurred at position of 225° for the $\beta = 0.35$ at the case of IPB-LC 3.

4.2 Effect of γ on the SCF Distribution

The parameter γ is the ratio of diameter or radius to wall thickness of the chord. In this study, there are three models with different of γ for each model, i.e. $\gamma = 10$, $\gamma = 12$ and $\gamma = 15$. Increase of γ in the models having constant value of chord diameter leads to decrease the chord wall thickness. All parameters except the γ parameter, i.e., the parameters β, τ and θ are made equal for the three models ($\beta = 0.44$, $\tau = 0.5$, and $\theta = 46°$).

Result of investigation can be seen in Figure 12 which is opposite to the β effect previously described, that increase of the γ leads to also increase SCF distribution along the brace-chord intersection line. Although the all load conditions caused the peak SCF majority occurs in the area around the point of crown 2, but each load condition yielded different SCF peaks at the range of 135° to 225°.

The peak SCF occurs almost precisely in crown 2 point (180°) for all γ values for the case of IPB-LC 2. For the case of IPB-LC 1 and LC 3, two SCF peaks appeared at positions of around 135° and 225° for all γ. For the previous case of loading, the first SCF peak (at 135°) is slightly larger than the second one (at 225°) for all three γ values. Meanwhile, for the latter case, the second SCF peak (at 225°) drastically became larger as γ increase. Generally, the smallest SCF of about less than 1.6 occurred at the range of 135° to 225° for all γ values at the case of IPB-LC 1.

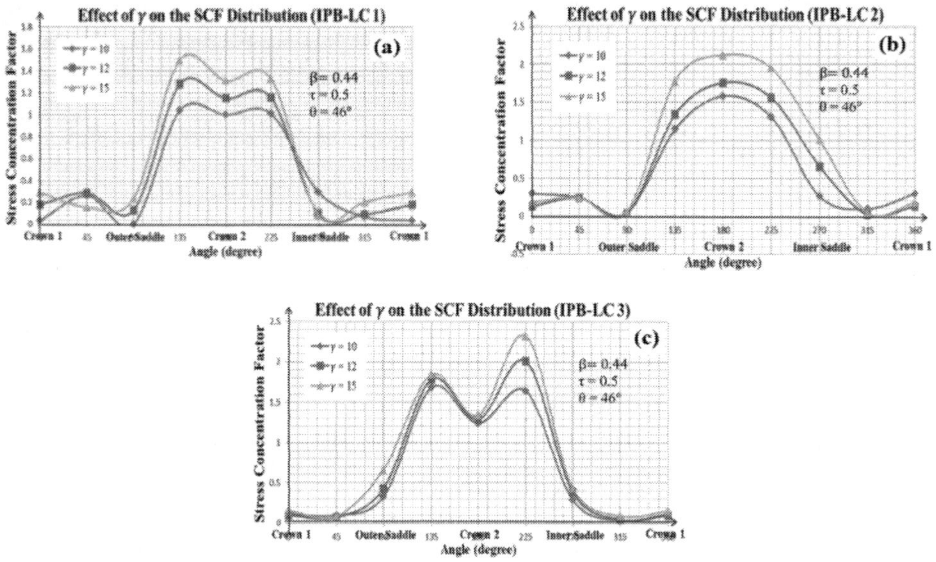

Fig. 12. Effect of γ on the SCF distribution due to in-plane bending loading: (a) Case of IPB-LC 1, (b) Case of IPB-LC 2, and (c) Case of IPB-LC 3.

4.3 Effect of τ on the SCF Distribution

The parameter τ is the ratio of brace to chord wall thickness. In this study, three models are used with different of τ for each model, i.e. $\tau = 0.35$, $\tau = 0.45$ and $\tau = 0.55$. Increase of τ in the models having constant value of chord wall thickness leads to increase of brace wall thickness. All parameters except the τ parameter, i.e., the parameters β, γ and θ are made equal to the three models ($\beta = 0.44$, $\gamma = 13.5$, and $\theta = 46°$).

As can be seen in Figure 13, effect of τ on the SCF distribution along the brace-chord intersection line almost similar to the γ effect previously described in terms of magnitude and pattern of the SCF distribution. Increase of the τ leads to also increase SCF, especially at the position range of 135° to 225° for all cases of IPB loading conditions.

Meanwhile, variation of the τ caused almost only slight changes to the SCF even though in the area around the point of crown 2 (at the range of 135° to 225°) for each load condition. The τ variation was not sensitive to the change of the SCF distribution along the brace-chord intersection line rather than the previous two parameters of β and γ.

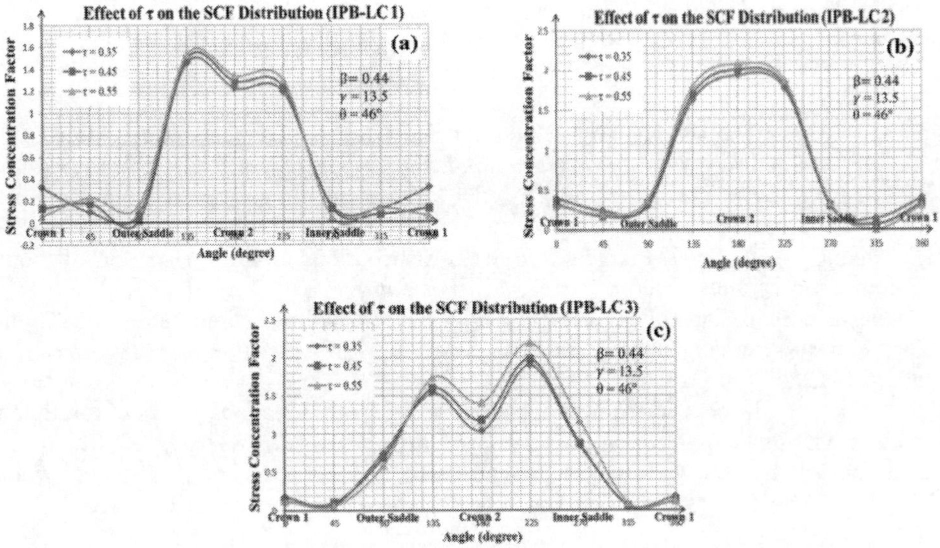

Fig. 13. Effect of τ on the SCF distribution due to in-plane bending loading: (a) Case of IPB-LC 1, (b) Case of IPB-LC 2, and (c) Case of IPB-LC 3.

Conclusions

Conclusions can be drawn from the present study are as follow:

- The results of parametric studies show that the peak SCFs mostly occurs at around Crown 2 point of the outer central brace at positions of around 135° and 225° for all three IPB loading conditions.
- The increase of the β leads to decrease the SCF. While the increase of the τ and γ leads to increase the SCF.
- The effect of parameter β and γ on the SCF distribution along the brace-chord intersection line are greater than the parameter τ even though in the area around the Crown 2 (at the range of 135° to 225°) for all three load conditions.

The authors wish to thank Institute for Research and Community Services (LPPM), Institut Teknologi Sepuluh Nopember (ITS) Surabaya for the financial support through a research grant scheme of “Laboratory Research 2017”.

References

1. Ahmadi, Hamid, E. Zavvar. Stress concentration factors induced by out-of-plane bending loads in ring-stiffened tubular KT-joints of jacket structures, *Thin Walled Structures*, Vol. **91**, 82-95, (2015)
2. U. Hidayat, R.W. Prastianto, W. Wardhana. SCF parametric study of multiplanar tubular joint double K. *Bachelor Thesis of Ocean Engineering Department FTK-ITS.* Surabaya. (2016)
3. D.S. Murthy. An Efficient FE Modelling Strategy for Fracture Analysis of Tubular

Joints. *Journal Structural Engineering Research Centre*, Vol. **85**, May, (2004)
4. Yeoh, Seng-Keat, Ai-Kah Soh and Chee-Kiong Soh. Behavior of Tubular T-Joints Subjected to Combined Loadings. *Journal Construction Steel Research*, Vol. 32, 259-280, (1995)
5. Soh, Ai-Kah, Chee-Kiong Soh. Hot Spot Stress of K Tubular Joints Subjected to Combined Loadings. *Journal Construction Steel Research*, Vol. **26**, 125-140, (1993)
6. H.L.J. Pang, C.W. Lee. Three-dimensional finite element analysis of a tubular T-joint under combined axial and bending loading. *Journal Fatigue*, Vol. **17**, 313-320, (1995)
7. E. Chang, W.D. Dover. Predicted of stress distributions along intersection of tubular Y-joints and T-joints. *International Journal of Fatigue*, Vol. **21**, 361-381, (1999)
8. Young-Bo, Shao, Du Zhi-Fu, Lie Seng-Tjhen. Predicted of hot spot stress distribution for tubular K-joint under basic loadings. *Journal of Construction Steel Research*, Vol. **65**. 2011-2026, (2009)
9. S.P. Chiew, C.K. Soh, N.W. Wu. General SCF design equations for steel multiplanar tubular XX-joints. *International Journal of Fatigue*, Vol. **22**, 283-293, (2000)
10. Karamos, A. Spyros, A. Romejin, J. Wardenier. SCF equations in multi-planar welded tubular DT-joint including bending effects. *Marine Structures*, Vol. **15**, 157-173, (2002)
11. Wingerde, A.M. Van, J.A. Packer, J. Wardenier. Simplified SCF formulae and graphs for CHS and RHS K- and KK-connections. *Journal of Construction Steel Research*, Vol. **57**. 221-252, (2001)
12. C.O. Woghiren, F.P. Brennan. Weld toe stress concentrations in multi-planar stiffened tubular KK-joints. *International Journal of Fatigue*, Vol. **31**, 164-172, (2009)
13. B. Naibaho, R.W. Prastianto, W. Wardhana. SCF parametric study of multiplanar tubular KDT-joint. *Bachelor Thesis of Ocean Engineering Department FTK-ITS.* Surabaya, (2016)
14. A. Puspaningtyas, R.W. Prastianto, W. Wardhana. Parametric study of stress intensity factor on multiplanar tubular joint with fracture mechanics. *Bachelor Thesis of Ocean Engineering Department FTK-ITS.* Surabaya, (2016)
15. I.F. Fuadi, R.W. Prastianto, N. Syahroni. Formulation of stress concentration factors for multiplanar tubular DKT-joint using finite element method. *Bachelor Thesis of Ocean Engineering Department FTK-ITS.* Surabaya, (2017)
16. Structural welding code: AWS D 1.1. USA, American Welding Society (AWS), (2002)
17. DNVGL RP-C203. Recommended Practice, Fatigue Strength Analysis of Offshore Steel Structure, (2001)
18. M. Haghpanahi, H. Pirali. Hot Spot Stress Determination for a Tubular T-Joint under Combined Axial and Bending Loading. *International Journal of Engineering Sciences*, Vol. **17**, 21-28, (2006)

MATEC Web of Conferences **177**, 01007 (2018) https://doi.org/10.1051/matecconf/201817701007
ISOCEEN 2017

Comparative Study of Structural Geometric to the Ultimate Strength on Fixed Jacket Platform

Muhammad Zubair Muis Alie[1,*], *Icon Yan Franchover* Obednego[1] , *Achmad Yasir* Baeda[1], *Taufiqur* Rachman[1], and *Juswan*[1]

[1]Hasanuddin University, Engineering Faculty, Department of Ocean Engineering, 92172 Gowa, South Sulawesi, Indonesia

Abstract. The brace configuration plays an important role to the ultimate strength of fixed jacket platform. The braces against the combined load in vertical and horizontal direction. In the present study, the ultimate strength of the fixed jacket platform is analysed considering the shape of the structural geometric. Four types of brace configuration namely, K, N, X and Y are taken to be assessed. Dimensions of the structure are constant including properties and materials. The boundary conditions are assumed to be fixed at the bottom part. The Non-Linear Finite Element Analysis (NLFEA) is adopted to calculate the ultimate strength of the structure and those results of brace configuration are compared with one another and discussed in the present study.

1 Introduction

Jacket platform is one of the common offshore platforms which are used at shallow and middle water. Generally, the structure has three to eight of leg because it has a good stability against the axial and lateral loads. The structure consists of three components; those are deck, jacket and pile. Deck is structural component located above water line, while jacket and pile are submerged into the water. The supported structure called jacket which functioned for exploration and other activities.

The ultimate strength has been assessed by many researches. The configuration effect of the symmetrical and unsymmetrical shape tubular member in buckling strength on fixed offshore platform has been discussed by [1]. Two kinds of offshore structures were taken to be considered in the analysis. The axial and lateral loads were imposed to jacket legs and other structural components. The material and dimensions were assumed to be constant and homogeneous. The boundary conditions were idealized to be fixed at bottom level. To assess buckling strength, the Finite Element Method (FEM) was used. Buckling and fatigue strength were assessed for jacket structures under symmetrical and unsymmetrical configuration shape. The Finite Element Method (FEM) was adopted to calculate those structures. The result obtained

* Corresponding author: zubair.m@eng.unhas.ac.id

by Finite Element Analysis was compared with the analytical solution. In addition, the Response Amplitude Operator (RAO) was applied to obtain fatigue life [2]. A closed-form prediction of buckling stress in local mode, including interaction of connected elements, and distortional mode, including consideration of the elastic and geometric stiffness at the web/flange juncture was conducted [3]. A mathematical model for slip buckling and its analytical solution has been founded for the analysis of layered and geometrically perfect composite columns with inter-layer slip between the layers was proposed [4]. Also, buckling of non-uniform column was analysed using Finite Element Analysis [5]. A research and development project for structural bamboo species were investigated [6].

In the present study, the ultimate strength is analysed using Finite Element Method considering the configuration shape on the jacket platform. Four types of braces namely K, N, X and Y. Dimensions of the structure are constant including properties and materials. The boundary conditions are assumed to be fixed at seabed. The results of brace configuration are compared with one another.

2 Finite Element Modelling

The finite element modelling of those structures considering the shape difference of the brace configurations X, Y, K and N are described in Fig. 1. The material and section properties are set to be constant. Depth and dimension of those structures are also constant. The braces configurations of those structures are illustrated in Figure 1. It should be noted that the boundary conditions at the end support on the seabed are fixed.

(a) Brace X

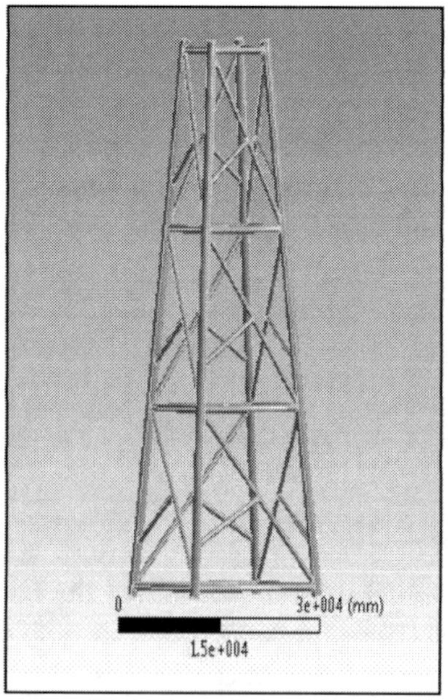

(b) Brace Y

(c) Brace K

(d) Brace N

Fig. 1. Braces Configurations of Fixed Offshore Platform

3 Results and Discussions

The deformation of those structures are shown in Fig. 2. It is generally observed that the maximum deformations are located near the lower part.

(a) Brace X

(b) Brace Y

(c) Brace K

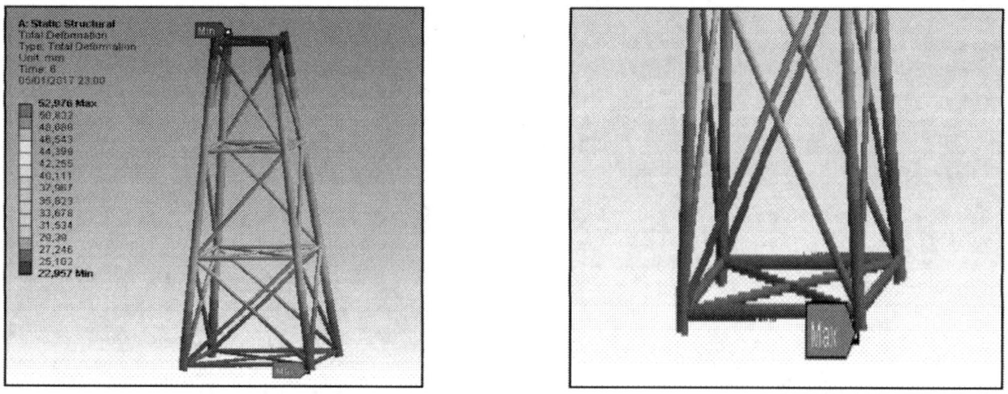

(d) Brace N

Fig. 2. Deformations of the Structure

The maximum deformations concentrate at local part as illustrated in Fig. 2. It should be noted that the end support of the jacket legs are assumed to be fixed. The stress-strain curve relationship obtained by Finite Element Analysis for four type of fixed offshore platform are shown in Fig.3. The axial force is applied to those structure and it is distributed to all jacket legs. The stress-strain relationship of each structure with brace configuration X, Y, K and N are identical in elastic region. For plastic, the stress-strain relationship for brace-Y is significantly increased compare to brace K, N and X. The stress-strain curve of brace-Y may be caused by the configuration of the structural braces above of the Brace-X component at the lower part. It is known that the Brace-X at lower part of those structures are identical. However, the above configurations of those structures are totally different with one another. The stress-strain curve of brace-X of the structure is smaller, than followed by Brace-N, Brace-K and Brace-Y. This may due to the brace-X is more rigid for the structural component when the structure againts the vertical loading even the horizontal loading is applied to the structure. According to the result shows that the brace-Y gives larger stress compares to the others which is better for design purpose.

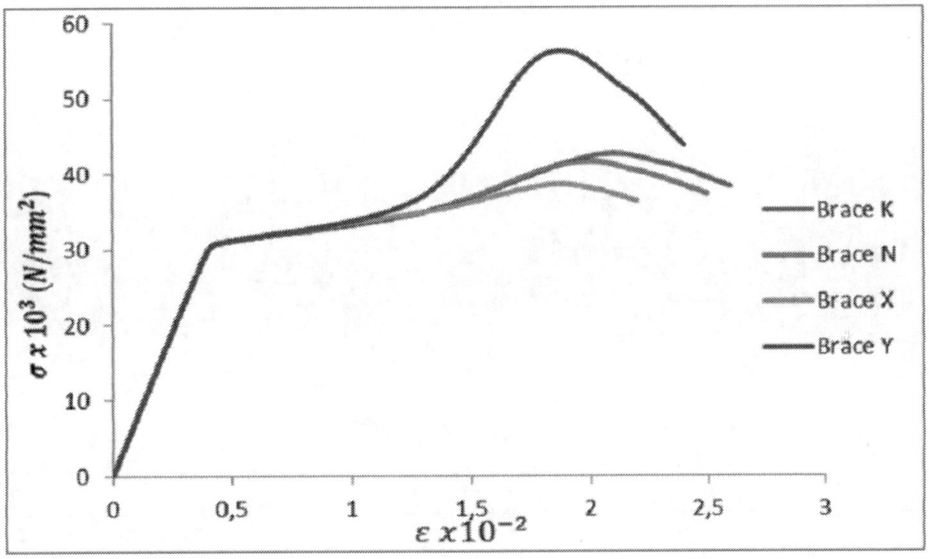

Fig. 3. Stress-strain Curve Relationship

4 Conclusions

The ultimate strength of brace-Y is significant difference compare to X, K and N braces. This phenomenon indicated by the stress-strain relationship of those structures. The characteristics of those structure both deformations and stress-strain relationship are caused by the configuration shape of the braces located at the middle and upper part.

References

1. Muis Alie, M.Z, Juswan, Daud, S, Sriadi, W.T, Ratnawati. Proceeding ISOPE. 1365-1371, (2015).
2. Muis Alie, M.Z, Int. J. of Technology, 1107-1116, (2016).
3. Schafer, B. W, J. of Structural Engineering, 289-299, (2002)
4. Kryzanowski, A et al, Int. J. of Solids and Structure, 2929-2938 (2009).
5. Iremonger, M.J, J. Computer and Structures, 741-748 (1980).
6. Yu, W.K et al, J. of Engineering Structures, 755-768 (2003).

Coastal Monitoring Program of Bathymetry Using Video Images Technique

Muhammad Zikra[1,*], *Noriaki* Hashimoto[2]

[1]Department of Ocean Engineering, Institut Teknologi Sepuluh Nopember, Surabaya, Indonesia
[2]Faculty of Engineering, Kyushu University, Fukuoka, Japan

Abstract. The objective of this research is to investigate the capabilities of optical remote sensing to monitor bathymetry in nearshore area using video images. The technique is designed to extract wave number components based on variation of intensity of brightness at each pixel in the images using cross-spectral correlation approach. This approach is based on pixel array analysis that utilizes a nonlinear inverse method 'Levenberg-Marquardt'. The technique is applied to the data collected at Hasaki beach in Japan from August to December 2006. The results indicate that the estimate of nearshore bathymetry is proved reasonable accurate near shoreline and breaking area where the differences between estimated and survey water depth are less than 10-30 cm.

1 Introduction

Bathymetry information is very important for coastal and marine engineers to understand the coastal process in nearshore area. Some nearshore activities such as recreation, fishing, navigation, beach nourishment and dredging require the knowledge of bathymetry. Good quality bathymetry information is hence required in order to make clear the physical processes that are taking place. However, combination of traditional in situ survey methods and advanced techniques such as global positioning system and modern ship vehicles are time consuming, and high cost, especially in shallow coastal water area.

The various remote sensing techniques have been applied to derive bathymetry estimate from images. These include measurement of water depth from aerial photograph [1], synthetic aperture radar (SAR) [2], X band radar [3] and SPOT (satellite) images [4] where the sequence of these images can provide a way to record information about changes in seabed topography. Recently, the invention of new digital technology of images form video camera system now can provide and improve an additional capability of automated data collection [5]. This automated data collections have much greater range of time and spatial scales. Also, this technology is suitable for measuring hazardous coastal areas such as surf zone area, where the operations of ship vehicle have limitation on maneuvers.

In these video image data it is possible to see the interaction of the incident wave field with the bathymetry (i.e. wave shoaling and refraction); hence this information can be used to obtain estimates of bathymetry [6][7]. The approach for estimating bathymetry is based on

* Corresponding author: mzikro@oe.its.ac.id

wave kinematics utilize the depth dependence of the wave speed (or, equivalently, the wavelength and frequency via c=f/k, where c is the wave phase speed, f is the wave frequency and k is the wave number = 1/L, and L is the wavelength). Overall, this approach requires image sequence (or time series of brightness intensity at discretely sampled locations).

Thus, the objective of this research is to investigate the capabilities of optical remote sensing to monitor bathymetry in the nearshore area using video images. The technique is designed to extract wave number components based on variation of brightness intensity at each pixel in the images using cross-spectral correlation approach. This approach based on the pixel array analysis utilizes a nonlinear inverse method 'Levenberg-Marquardt' to find the optimal wavenumber estimate from a model with the observation data set. In this research, the video images measured at Hasaki beach in Ibaragi prefecture in Japan is used to investigate the applicability of the model. In the following sections, we will first review data conditions measured at Hasaki beach area. Next, we summarize mathematical formation of wave number model and inversion method. Finally, we apply the methods to the data, and then draw some conclusions.

2 Methodology

In this research, observation was carried out with video camera at Hasaki beach, Japan. The Hasaki beach is located on 120 km east of Tokyo facing the North Pacific Ocean as shown in Figure 1. In general, Hasaki beach is known as straight sandy coast stretching from north to south with length around 17 km long. Since 1986, many coastal studies have been conducted in this location especially around the pier which is known as HORS (Hasaki Oceanographical Research Station).

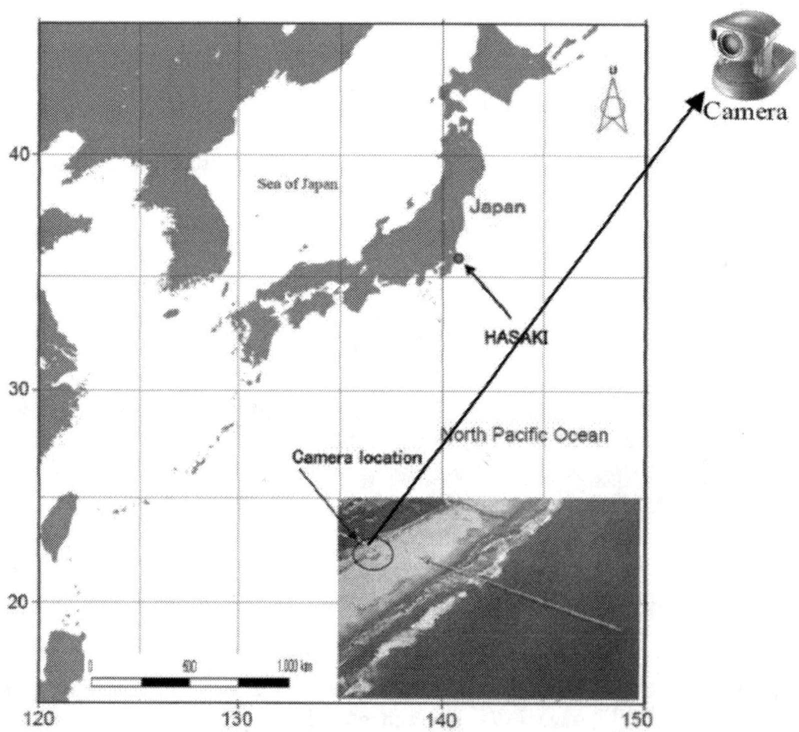

Fig. 1 Study Location

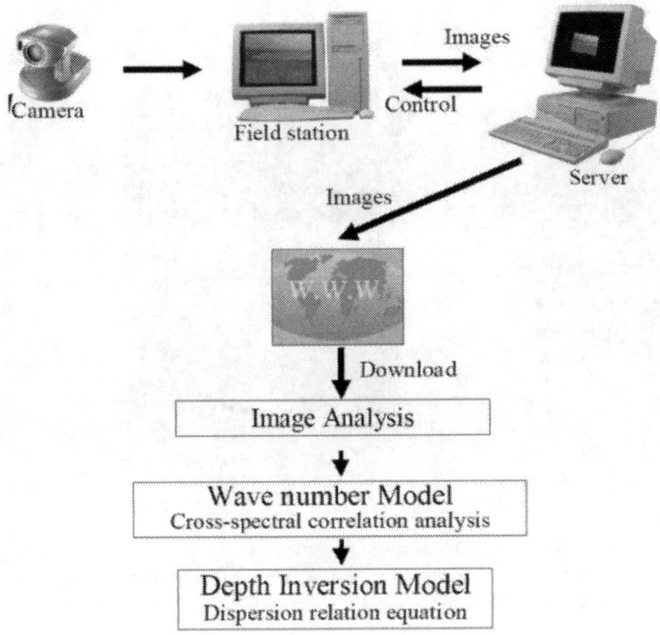

Fig. 2 Methodology of bathymetry monitoring using video images

2.1 Video Camera System

The video camera system in Hasaki site was first installed on August 16, 2006. The data sets of video images were collected from single camera network Canon VB-C50iR on 10 m height above the ground level to generate images with the resolution of 640 x 480 pixels. In this video camera system, snapshot images were collected at interval 1 second every hours using single camera as shown in Figure 2.

2.2 Image Analysis

In the image analysis, the procedures consist of image rectification and timestack image analysis in which the main aim is to obtain physical information from timestack images. In the Hasaki site, the camera took successive snapshot images around Hasaki pier at the interval of 1 second. To extract pixel lines from successive snapshot images, the image coordinate of pixel (u,v) on the snapshot image need to convert on the real coordinate system (x,y,z). In this rectification, the relationship between image coordinate and real coordinate as described by Holland et al, 1997 was used. The result of rectification image from snapshot image is shown in Figure 3 (middle image).

Then timestack images were collected hourly at each point in the array which can be expressed as I (xi,yi,t), where xi,yi are the spatial coordinate of the ith image pixel, and t is discrete sampling time. An example of time series of brightness intensitiy of pixel for cross-shore at y = 120 m is presented in Figure 2 (right image).

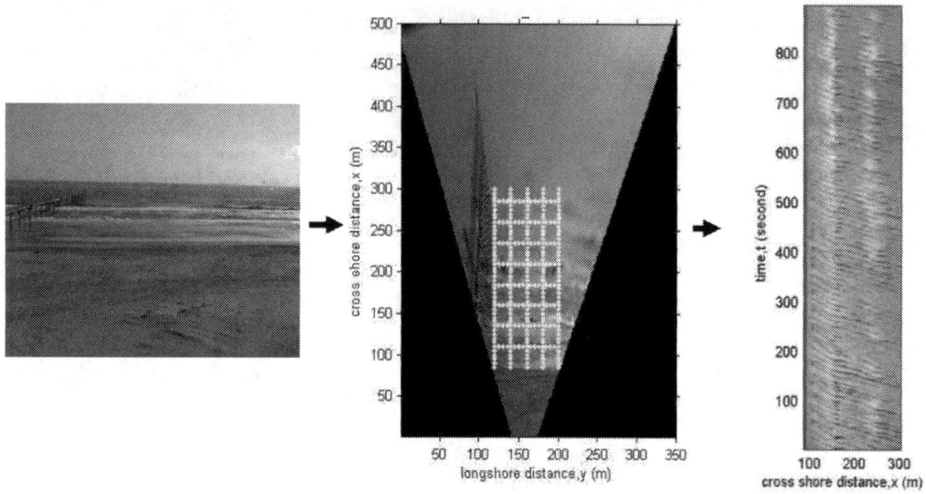

Fig. 3 Rectified image from snapshot image on 25/08/2006 at 07.00 around pier area with five cross-shore arrays and eight long-shore arrays (shown with dots pixel)

2.3 Inversion Method

In the present research we are interested in the estimation of nearshore bathymetry from video images. The technique is based on wave number component where it can be derived from the intensity of the brightness at each pixel in the images by using cross-spectral correlation approach [8].

The mathematical formulation for cross-spectral correlation model between two pixels is described by

$$C_{i,j,f}^{MODEL} = \exp\left\{2\pi\Delta x\sqrt{-1}\sum_{m=1}^{M} D_{i,j,m}k_{m,f}\cos\left(\alpha_{m,f}\right)\right\} \tag{1}$$

where f is wave frequency, Δx is spacing between pixels, D is design matrix defined on both sample domain (xi, xj), α (wave direction) and k (wave number) as unknown model parameters.

The sample design matrix, D is designed as basis function:

$$D_{i,j,m} = \sum_{i'=i}^{j} a_{i,m} \tag{2}$$

where a_i, m is smoothing weight of Hanning filter

$$a(r_{i,m}) = \left\{1-\cos\left(0.5\pi\left[1+r_{i,m}\right]\right)\right\}^2 \tag{3}$$

$$r_{i,m} = \left|x_i - x_m\right| L_x^{-1}$$

where L_x is smoothing lengths scale. Using observation data from timestack images, cross-spectral observation data can be acquired by applying discrete Fourier transform to the observation to compute the cross-spectra between two pixels (sensors pair) [9].

$$C_{i,j,f}^{OBS} = \left\langle \tilde{I}(x_i, f)\tilde{I}^*(x_j, f) \right\rangle \quad (4)$$

where the tildes indicate the Fourier transform, the asterisk indicates complex conjugate, angle brackets indicate ensemble or band averaging.

Since the wave number is nonlinearly related to the cross-spectral correlation as shown by equation above, a non-linear inversion method Levenberg-Marquardt [10] is used to minimize the weighted squared difference between successive estimates of the model and the observations.

$$\Delta C_{i,j,f}^{\tau} = \left\{ \gamma_{i,j,f} C_{i,j,f}^{MODEL(\tau)} - C_{i,j,f}^{OBS} \right\} \quad (5)$$

where, at each iteration τ, the model-observation mismatch is weighted by the observed coherence, $\gamma_{i,j,f}$.

The minimization of cross-spectral function with Levenberg-Marquardt (LM) method consists on constructing an iterative procedure that starts with an initial value guess k_o. Then, a new estimate of wave number model is obtained with:

$$k_{f,m}^{\tau+1} = k_{f,m}^{\tau} + \Delta k_{f,m}^{\tau} \quad (6)$$

where the variation of $\Delta k_{f,m}^{\tau}$ calculated from

$$\Delta k_{f,m}^{\tau} = \left(\left[R^{\tau} \right]^T R^{\tau} + \lambda^{\tau} I \right)^{-1} \left[R^{\tau} \right]^T \Delta C_{i,j,f}^{\tau} \quad (7)$$

where λ^{τ} is the damping parameter, I is the identity matrix and R is sensitivity matrix for the cross-spectral correlation as describe:

$$R^{\tau} = \gamma_{i,j,f} \sqrt{-1} D_{i,j,m} C_{i,j,f}^{MODEL(\tau)} \Delta x \quad (8)$$

The iteration procedure to calculate $\Delta k_{f,m}^{\tau}$ and $k_{f,m}^{\tau+1}$ from Eq. 6 and 7 is continued until the convergence criterion

$$\left|\Delta k_{f,m}^{\tau}\right| < \varepsilon \qquad (9)$$

is satisfied, where ε in this model is 10^{-6}

2.4 Bathymetry Inversion

The bathymetry inversion method based on timestack method computes water depth by relating wave number parameters k using a suitably accurate dispersion equation. Water depth h is related to local wave number k and frequency f through the dispersion relationship in the linear wave theory [13]

$$(2\pi f)^2 = gk\tanh(kh) \qquad (10)$$

where g is gravitational acceleration and h is local water depth.

Given a value for f (sample wave frequencies) and an initial depth, h, this equation can be solved iteratively for wave number. The Levenberg-Marquardt non-linear inverse method was used again to minimize error between the wave number predicted by Eq.10 and the wave number estimated form images derived from Eq.6.

3 Analysis and Results

This technique was applied using timestack data collected on August 25, 2006. Using this timestack, the wave number estimate was computed at a series of wave frequencies ranging from 0.08 Hz to 0.11 Hz. It is expected to find suitable frequency from those wave frequency resolutions; the optimum wave component will give strong signal for the analysis of brightness intensity of pixel time series. On this date, the peak wave period based on field measurement was 9.1 sec; the wave direction was approach from 81 degree from North direction; and the significant wave height was 1.11 m.

Using the measured bathymetry and the tidal level at the time of the image collection, the wave number estimate was computed for each frequency. The non-linear inversion method of Levenberg-Marquardt was applied to the sample cross-spectral correlation at each frequency over entire array. Figure 4 shows the results of the wave number estimates for each sample frequency. The best of wave number estimates were obtained at frequency 0.09 Hz which showed highest coherence with rms error 0.0342 m-1 as shown in Table 1. This frequency, 0.09 Hz corresponds with the peak period based on field measurement.

The result of bathymetry inversion using video images data collected on August 2006 is shown in Fig 6. The average bathymetry value estimated by inverted all frequency shows that the Hasaki beach have sand bar at x = 200-200 m. To evaluate bathymetry inversion model, the result is compare to the survey measurement data. It shows that the performance of the prediction is most accurate near shoreline and sand bar, where the differences between estimated water depth and survey water depth is less than 10-30 cm.

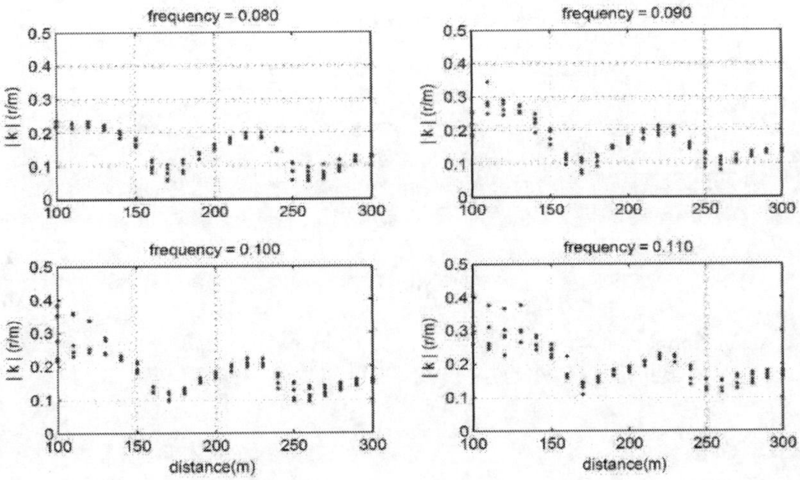

Fig. 4 Wave number estimates using cross-spectral correlation from snapshot image at 25/08/2006

Table 1. Coherence and RMS Error Wave number

No	Frequency	Coherence	RMS wave number error
1	0.08	128.745	0.0358
2	0.09	132.153	0.0342
3	0.10	131.161	0.0354
4	0.11	118.333	0.0421

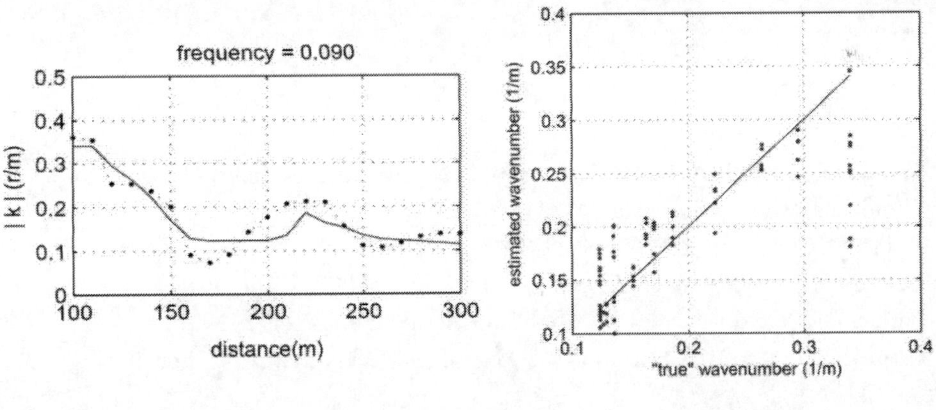

Fig. 5 (Left figure) Comparison of wave number estimated from video images (dot) and wave number from linear theory (line). Data represent from frequency 0.09 Hz. (Right figure) Correlation coefficient between wave number estimated and "true" wave number is 0.93.

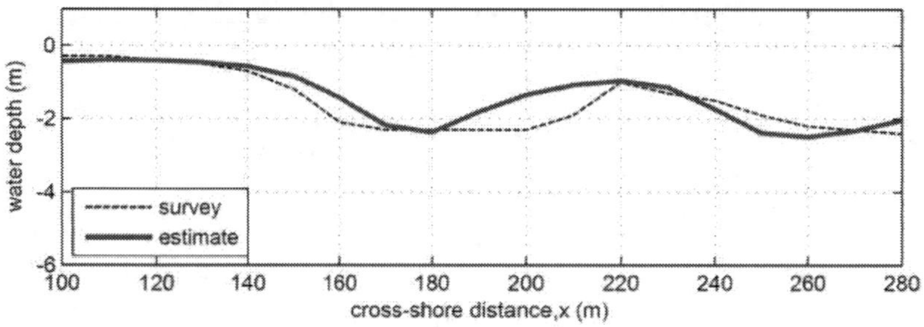

Fig. 6 Comparison between cross section profile in situ survey (dash line) and bathymetry inversion (solid line) from video images collected on August 2006.

4 Conclusions

An algorithm for video images sequence analysis was presented to monitor bathymetry in near shore area using non-linear inversion method to convert image pixel brightness values into depth estimates. The method consists of wave number inversion model which is based on cross-spectral correlation technique and bathymetry inversion which is based on wave dispersion model. The capability of video image technique was tested using data from Hasaki beach in Japan from August 2006 to December 2006. The results indicated that cross-spectral correlation approach have the capability to derive wave number estimate from time series of brightness intensity at each pixels to estimate bathymetry. The model showed relatively small rms errors between 0.0342-0.0421. Correlation coefficient between the estimated wave number and that from linear wave theory is 0.93.The result of nearshore bathymetry estimates provides reasonably accurate depth estimate near shoreline and breaking area with rms 0.44. This result indicates that nearshore bathymetry estimates can be derived from video images sequence.

References

1. Wiegel, R.L. and R.A. Fuchs. Rep. **74-9**, Univ. of Calif., Berkeley (1953)
2. Greidanus, H., Hydrographic j., **83**, 13-18 (1997)
3. Bell, PS., Coastal Eng, **37**, 513-527 (1999)
4. Leu, L.G., Y.Y Kuo, and C.T. Lui, Coastal Eng. J, **41**, 21-41 (1999)
5. Holman, R.A and J. Stanley, Coastal Engineering, **54**, 477-491 (1991).
6. Holman, R.A., T.C. Lippmann, P.V. O'Neill, and K. Hathaway, Marine Geology, **97**, 225-231 (1991)
7. Holland, K.T, R.A. Holman, T.C. Lipmann, J. Stanley and Plant, IEEE J. Oceanic Engineering, **22**(1), pp.81-92 (1997)
8. Zikra, M., Hashimoto, N., Yamashiro, M., & Suzuki, K. Spectral Analysis of Pixel Brightness on Video Images For Wave Analysis At Hasaki Beach, Japan. *Proceedings of the Twenty-first International Offshore and Polar Engineering Conference*, pp.974-979 (2011).
9. Zikra, M., Hashimoto, N., Yamashiro, M., Yokota, M. and Suzuki, K. Coastal Engineering Journal, **54** (03), 24 pp., DOI: 10.1142/S0578563412500209. (2012)

10. Bendat, J.S., and A.G. Piersol, "*Random Data: Analysis & Measurement Techniques*", 566 pp., Wiley Intersci., New York (2000).
11. Press, W.H, Teukolsky, S.A, Vetterling, W.T, and Flannery, B.P., "*Numerical Recipes in C: The Art of Scientific Computing", 2nd ed*, Cambridge University Press (1992).
12. Stockdon, H.F. and Holman, R.A., Journal of Geophysical Research **105**, pp. 22015-22033 (2000).
13. Dean, R.G., and R. A. Dalrymple, "*Water Waves Mechanics for Engineer and Scientist*", 353 pp., World Sci., River Edge, N.J (1991).

Experimental Study of Scouring Characteristic Around Hexagonal Artificial Reef

Haryo Dwito Armono[1,*] and *Harish* Wirayuhanto[2]

[1] Associate Professor, Ocean Engineering Department, Faculty of Marine Technology, Institut Teknologi Sepuluh Nopember, Surabaya, Indonesia

[2]Graduate Student, Ocean Engineering Department, Faculty of Marine Technology, Institut Teknologi Sepuluh Nopember, Surabaya, Indonesia

Abstract. Artificial reef is a structure placed underwater which has similar function with natural reef. Artificial reef can also serve as a submerged breakwater. When placed in shallow water, the structure is vulnerable to scouring and soil subsidence which cause structural failure. This paper present serial testings in Ocean Environment and Ocean Energy Laboratory, Ocean Engineering Department, ITS, to determine the scouring characteristic with variation of gap, and waves for 1: 10 model scale hexagonal artificial reef. The results of this research can guide the planning of structural protection against scouring. From a series of tests, it can be summarized that for 1 diameter gap reef distance, the maximum scouring occurs behind the structure, whereas for no gap distance, it occurs at the front of the structure. At the center of the structure, for no gap distance, sedimentation occurs dominantly, whereas for 1 diameter gap distance, scouring process occurs dominantly. The maximum scouring depth (5,3cm) occurs when the gap is equal to 1 diameter of reefs and minimum scour(2,9 cm) occurs when there is no gap between the reef.

1 Introduction

The damage of coral reefs affects the lives of many kinds of marine organisms, especially in tropical waters where the marine productscould decline severely due to low productivity of coral reefs. Coral reef recondition needs to be taken to restore function and productivities. One effort that can be done to restore the condition of coral reefs is to transplant coral on artificial reefs. Artificial reef are objects or structures placed on the sea floor that expected to resemble the characteristics natural coral reef. This structure is used to provide a sanctuary for marine animals and serves as a submerged breakwater that reduces wave energy came before the shoreline [1]. There are several types of artificial reefs that have been developed, such as Aqua Reef, Turtle Reef, Reef Ball, Ultra Ball, Bay Ball, Square Reef and Reef Bottle. In this study, the new reef model is developed in the form of hexagon called Hexareef.

Any submerged breakwater, or in this case Hexareef and other types of structures when placed on a marine environment with a dynamic water flow, will lead to changes in water flow field, thus forming a vortex flow around the structures that cause scouring [2,3]. If the structure scouring occurs continuously, the foundation is unstable,and lead

*Corresponding author :armono@oe.its.ac.id

to structural failure. Failure of structures include overturning, settling, and sliding[4].Scouring due to wave caused by the difference of hydostatic pressure on the ocean floor following the fluctuation of surface waves. Hydrostatic pressure is transmitted to the seabed and formed a huge difference in the vertical and horizontal pressure that causes soil liquefaction and scouring at the seabed [5,6]. The shape of the struture affect the current flow and waves that cause scouring. The depth of scouring that occurs around the square and hexagonal structure with the placement of wave direction different is greater than the depth of scouring around the cylindrical structure [5].

Furthermore, scouring depth will affect some of the factors that cause changes in water conditions including wave height, wave period, water depth, and the type of waves generated [7,8]. These factors are summarizedintoanon-dimensional parameter that contains the independent variables in testing of the characteristics of scouring nearHexareef. These parameters include the number Keulegan Carpenter (KC) and mobility parameters on the seabed (Ψ) [7,9].

In this research, the pattern and characteristics of scouring that occurs in near Hexareef model affected by some wave parameters and variation of distance (gap) between Hexareef during test will be examined. The maximum and minimum scour will be noted, for Hexareef placement with and withoud gap. **Fig. 1**shows typical shape and arrangement of Hexareef during test.

Fig. 1. Hexagonal Artificial Reef Structures (Hexareef) Placement in the Wave Flume

2 Material and Methods

2.1 Testing Preparation

Physical model testing performed in the wave tank located on Laboratory of Marine Environmental and Marine Energy, Department of Ocean Engineering, Institut Teknologi Sepulluh Nopember (ITS) Surabaya. The installed wave generator has the capability of generating regular and irregular waves with various wave spectrums. The length of the wave tank is 24m with the height and width of 2m. The model placed in the middle of the wave tank, on top of35cm sand layer with the front slope of 1:7. The model were submerged on 65cm of water level. The tests performed for hexagonal reef with gap distance and variation period and wave height.

Due to limitation of quantity, only 2-layer Hexareef model arranged laterally will be tested.Therefore,a partition wall was installed in the middle of wave tank and only 57cm of wave tank used for the test. Figure 2 shows the arrangements of the models and wave probe placement in the wave tank during test.

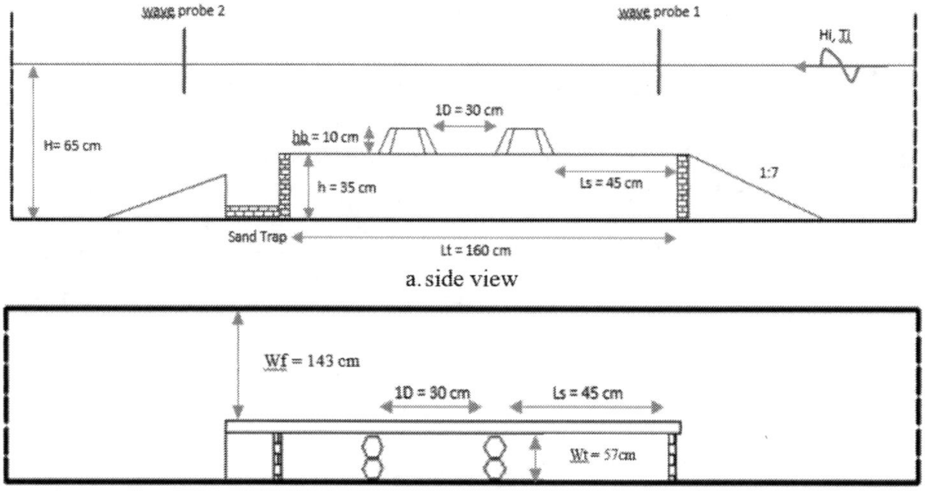

b. top view

Fig. 2. Testing plan in the wave flume

The models were made of concrete with the scale on 1: 10 to the prototype. With the height of 10 cm, the width of model was 18,5 cm at the top and 25 cm at the base The Hexareef models were placed in the sand layer with average diameter (D_{50}) of 0.28 mm and density of 2650 kg/m^3. The thickness of sand layer was 35 cm. The experiment will compare the scour between structure arrangement with 1 diameter (1D) to the structure without gap (0D).

Table 1. Number of Test,Wave properties, and Structure configuration

Regular Wave			
No. Test	Ti [sec]	Hi [m]	Gap
1	1	0,1	0D
2	1	0,2	0D
3	2	0,1	0D
4	2	0,2	0D
5	1	0,1	1D
6	1	0,2	1D
7	2	0,1	1D
8	2	0,2	1D

Irregular Wave JONSWAP)			
No. Test	Ti [sec]	Hi [m]	Gap
9	1	0,03	0D
10	1	0,05	0D
11	1	0,07	0D
12	1	0,03	1D
13	1	0,05	1D
14	1	0,07	1D

A series of reguler and irregular wave based on JONSWAP spectrum will be applied to Hexareef model as listed in Tabel 1. During tests, the variation not only made for wave height and period, but also distance (gap) between the structures. Initially the surface of sand was smoothed and levelled. The waves were generated for 20 minutes for each test. Every 5 minutes the waves generator were stopped, and the pattern on the sand surface were mapped and measured using laser distance meter.

2.2 Data Acquisition

A probe is installed in the front of the model to record the incoming waves. Another probes were placed behind the model to record the transmitted waves. The comparison of incoming and transmitted waves will not be discussed in this paper. However, the transmission coefficients were similar to other test of artificial reefs [1]. The fluctuation of water level recorded by the probes were analyzed to estimate the significant wave height and period.Callibration of each probes were made before the test to ensure the correct recording of water level fluctuation.Only maximum and minimum scour were noted for the whole test.

The pattern of depth of scour in the vicinity of the Hexareef model were measured and mapped every 5 minutes using laser distance meter. A map of sand contour before and after test were made every 20 minutes to estimate the maximum depth of scour and its location. A non-dimensional parameter will be developped to evaluate the influence of gap and wave characteristics to the depth and location of scour. Two locations will be examined during the test; in the front and rear of Hexareef models.

3 Result and Discussion

3.1 Waves Data

For every test performed, the water fluctuation level in the front probe will be recorded on Channel 1, while the water level fluctuation in the rear of the model were recorded on Channel 2. The recording data in this channel will be evaluated for its signficant wave height and period for each test and listed in Tabel 2.

Table 2. Generated Wave Height and Period for Each Test.

Regular Wave				
No. Test	Ch1(Front)		Ch2 (Rear)	
	Hm [cm]	T [sec]	Hm [cm]	T [sec]
1	8,3	2,0	12,1	2,0
2	17,2	2,0	23,1	2,0
3	11,4	4,0	11,1	4,0
4	17,3	4,0	21,8	4,0
5	9,7	2,0	12,7	2,0
6	19,5	2,0	23,5	2,0
7	11,3	4,0	10,5	4,0
8	18,2	4,0	21,7	4,0

Irregular Wave JONSWAP)				
No. Test	Ch1(Front)		Ch2 (Rear)	
	Hm [cm]	T [sec]	Hm [cm]	T [sec]
9	6,1	2,2	6,9	2,2
10	9,0	2,2	10,0	2,1
11	11,2	2,2	12,9	2,2
12	6,0	1,9	6,9	2,2
13	8,9	2,1	10,4	2,2
14	10,7	2,0	12,5	2,1

3.2 Scouring on Regular Waves Test

For the tests with similar wave height (Test 2 and Test 4), it was found that the larger the wave period, the greater the scouring that occurs in the structure. While test using wave with the same period (Test 3 and Test 4), the greater the incoming wave height, the greater the scouring that occurs.Based on experiments for regular wave above, for experiment without gap (0D), the biggest scour observed during Test 4 (17.3 cm wave height and wave period of 4 seconds). The elevation of sand layers changed gradually and undergoing expansion on the front and rear of the structure. In the end of test without gap, its observed that the front left of the model suffered sedimentation while scouring occurs in the back side of the model. At the center of the structure there was small sedimentation, but in certain parts experiencing scouring.

When the gap introduced in the model, most of the area suffer scouring. Small sedimentation and scour were observed inthe gap between structures. Overall, the depth of scour in the model with gap is bigger than model without gap. The maximum depth observed at Test 8 (18.2 cm wave height and wave period 4 seconds).Similar pattern to model with gap were found, that the larger the wave period, the greater the scouring that occurs in the surrounding area of the structure. While on wave with the same period (Test 7 and Test 8), the greater the incoming wave height, the greater the scouring that occurs. Due to limited space in this paper, the scouring patterns are not presented.

3.3 Scouring on Irregular Waves Test

During experiments with irregular waves, the biggest sand level change occurred in Test 11 (Hs 11.2 cm and Tp of 2.2 seconds) for model without gap (0D). While the test models with 1 diameter of gap, the biggest sand level change observed on Test 14 (Hs 10.7 cm and Tp 2 seconds). It is observed that the sand level changes mostly occurred at the rear of the structure. Several places at the rear and front of the structure experiencing sedimentation. In Test 14 where the gap is introduced, several locations on the front and rear structures also experiencing sedimentation, such as on the left front and the rear right of the structure. At the center of the structure (gap area), sedimentation only occurs on the right side of the structure, while on the left side of the structure scouring sand were observed.

Based on the regular and irregular test results above, the maximum scouring occurs on regular wave test. This is due to the amount of wave energy received by the sand layer is constant in regular waves, while in irregular wave, the energy fluctuates following the incoming waves.

3.4 Maximum Scouring Analysis

The maximum scouring depth (Sm) will be observed in this study. The maximum scouring depth is the maximum depth of sand level in the front and back of the Hexareef model after being tested in the wave tank for 60 minutes with an interval of recording per 5 minutes [7]. The maximum scouring depth (Sm), in the test without gap (0D)were in the range of 5 cm to 2.9 cm. For the test with gap distance 1D, range of maximum scouring depth were 5.3 cm to 3 cm. Table 3 below shows the maximum scouring depth recorded during experiment.

Table 3. Record of Maximum Scouring Depth for Each Test

Wave type	Gap	Hi [cm]	Ti [sec]	Scouring depth [cm]		W [cm]
				Front	Rear	
Regular	0D	8,3	2,0	2,9	3,7	37
		17,2	2,0	3,8	3,9	37
		11,4	4,0	4,4	4,2	37
		17,3	4,0	5,0	4,4	37
	1D	9,7	2,0	3,5	4,0	30,83
		19,5	2,0	4,3	4,0	30,83
		11,3	4,0	4,1	4,3	30,83
		18,2	4,0	4,2	5,3	30,83
Irregular (JONSWAP)	0D	6,1	2,2	3,6	3,5	37
		9,0	2,2	3,9	3,4	37
		11,2	2,2	4,0	3,6	37
	1D	6,0	1,9	3,0	3,0	30,83
		8,9	2,1	3,3	3,1	30,83
		10,7	2,0	3,6	3,1	30,83

In the Table 3 above, W is the effective width of the structure, and can be calculated using following expression

$$W = nA - (B/C)\,A \tag{1}$$

where n is the number of Hexareef model, the width of A, B and C as explain in Figure 3 below.

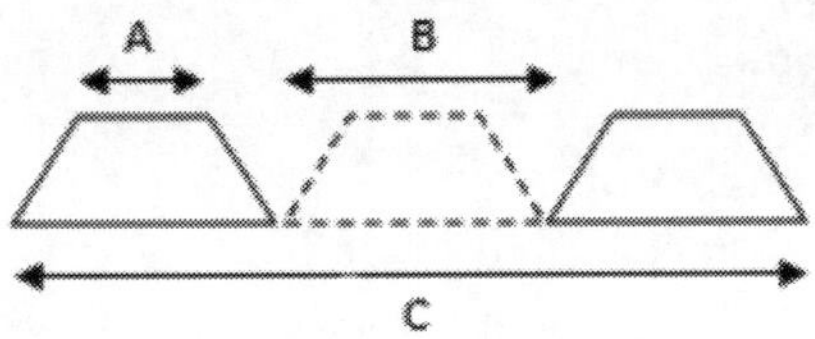

Fig. 3. Model Width Definition

The influence of wave height and period to maximum scouring depth (Sm) in the front and rear of model were shown in figure 4 below.

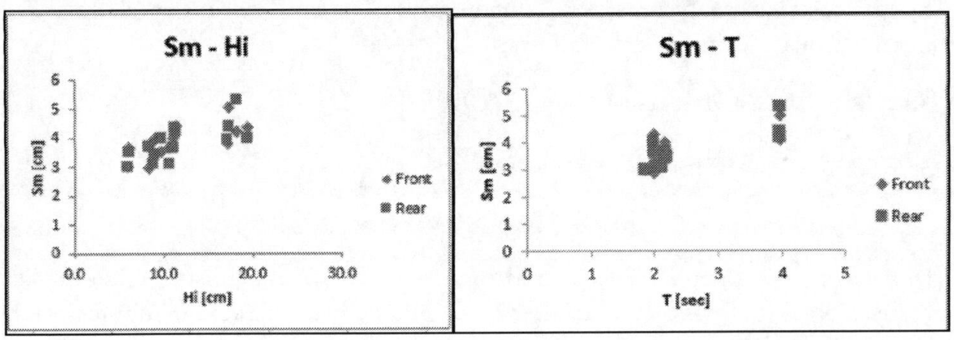

Fig. 4. The influence of wave height (Hi and wave period (T) to maximum scouring depth (Sm)

It isobserved that the greater the value of incoming waves height and period, the greater the maximum scouring depth occurred in the front and rear of the structure. Other variables also influence the maximum scour depth, such as effective width (W) and water depth (h) as shown in Figure 5 below.

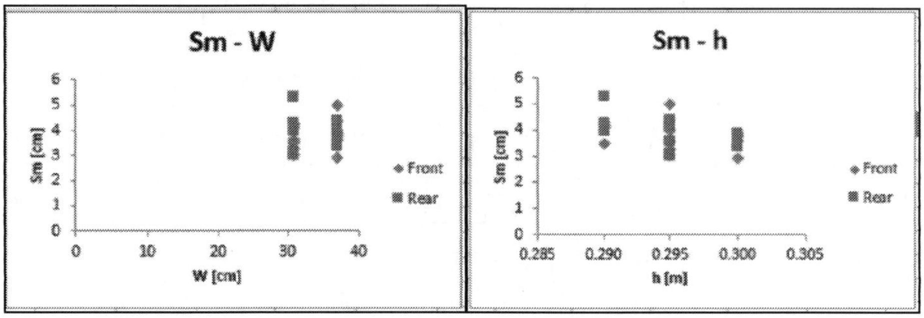

Fig.5. The influence of effective width (W) and water depth (h) to maximum scouring depth (Sm).

The smaller the effective width of the structure, the greater the scouring that occurs at the front and rear of the structure. The shallower the depth of water in an experiment, the greater the scouring that occurs at the front and rear of the structure.The influence of incoming wave height and period, effective width, water depth and other variables to maximum depth of scouring can be expressed as following equation:

$$Sm = f(W, Hi, T, h, d_{50}, \rho, \rho_s, g) \quad (2)$$

with,

S_m = maximum scouring depth [m]
W = effective width of structure [m]
H_i = incident wave height [m]
T = incident wave period [second]
h = water depth [m]
d_{50} = sediment mean diameter [m]
g = gravity acceleration [m/s^2]
ρ = water density [Kg/m^3]
ρ_s = sediment density [Kg/m^3]

By dimensional analysis, several dimensionless parameterscan be defined from equation 1 above and expresses as follows:

$$Sm / W = f(Hi/W, h/W, d_{50}/W, \rho_s/\rho, gT^2/W) \quad (3)$$

after compounding, equation (3) can be expressed as

$$Sm / W = f(Hi/W, h/W, d_{50}/W, \rho_s/\rho, (Hi/T)^2/(gh)) \quad (3)$$

Dimensionless parameter that is used to analyze the characteristics of scouring on submerged breakwater is Keulegan-Carpenter number and parameter number seabed mobility [9].

$$KC = H_i / W \quad (4)$$

$$\Psi = (H_i / T \sinh(kh))^2 / (g^*h) \quad (5)$$

By componding, those two dimensionless parametersabove can be developed from dimensionless parameters in Equation 3, making it suitable for use in the analysis of the characteristics of scouring. The main parameters of non-dimensional relationship between the depth of maximum scouring and the effective width of the structure (Sm/W) is influenced by variables making up the parameters of non-dimensional Keulegan-Carpenter (KC) by the square root of the parameter non-dimensional mobility number (Ψ), which form the non-dimensionalparameter $KC\Psi^{0.5}$ [9]

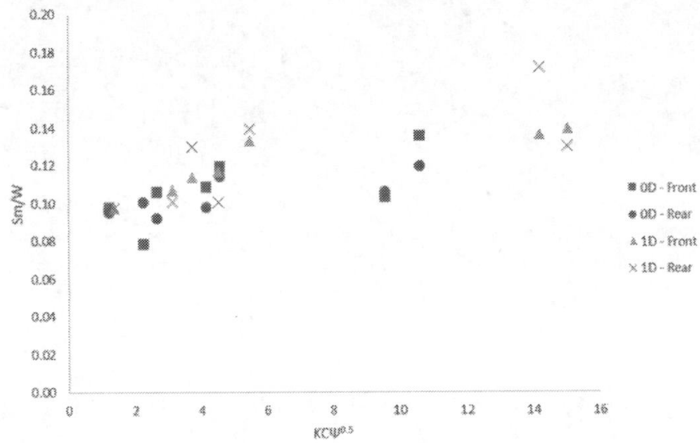

Fig. 6. Influence of KC number to maximum scour depth

Furthermore, following [2], a non-dimensionless parameter $KC\Psi^{0.5}$ were plotted againts ratio of maximum scouring depth to effective width as shown in Figure 6 above. Some note about the distribution of the maximum scouring depth can be observed in Figure 6, including:

1. The maximum scouring occurs in the location behind the model for experiment with 1D gap
2. The minimum value for the overall testing of scouring depth occurred in front of the structure for 0D gap.
3. In general, the maximum scouring occurs at 1D gap and the minimum at 0D gap

The maximum scouring depth for the front and rear structures can be estimated with an empirical equation developed by regression analysis [10] of the measured maximum scouring depth data obtained during the test. The empirical equations for the front and rear of Hexareef model expressed in equations 6 and 7 as follow.

$$S_m / W_{bw} = 0{,}0029(\psi^{0,5}KC) + 0.0963 \quad (6)$$

$$S_m / W_{bw} = 0{,}0035(\psi^{0,5}KC) + 0.0933 \quad (7)$$

From the results of the regression analysis above, (6) and (7) has a regression coefficient (R^2) 0.5713 and 0.4965. It shows that only 57.13% and 49.65% of independent variables ($KC\Psi^{0.5}$) had correlation to the dependent variable (Sm / W).

Several studies have been conducted on scouring evaluation around submerged breakwater such as [7] and [9],and empirical equations has been proposed from the study [9]

$$S_m / W_{bw} = 0{,}0125(\psi^{0,5}KC) \quad (7)$$

Figure 7 below is a comparison of the value of the maximum scouring depth on the rear of the structure of the research done by [7] and [9].

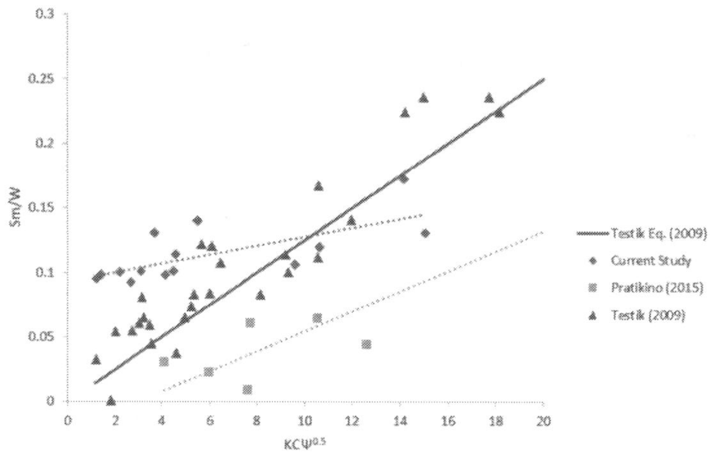

Fig. 7. Comparison withother Studies

The maximum scouring depth in this study has the increasing trend compared with other studies, which is the greater the value incoming wave height (Hi), the greater the depth of scouring. In this study, the scouring depth had a bigger value for the same wave height Hi and KC number compared to the research conducted by [7] and [9].

However, at certain value of KC number, the value of the maximum scouring depth in this study is smaller than research by [2] but larger than the research by [7]. Factor geometry of the structure, the depth of the waters of the test area, and the diameter of the granules also affects the maximum scouring depth. Geometry factors affect the rate of current flow formed around a structure, this will lead the scouring processes.

4 Conclusion

The following conclusion can be drawn from the study on the characteristics of scouring around Hexareef:

1. The higher of wave height and period resulted in the deeper scour that occurs around the Hexareef. The maximum scouring depth occurred in the range of 5.3 cm to 2.9 cm. The maximum occurs on the test using regular waves.
2. The smaller the gap structure, the smaller scouring that occurs, while the larger the gap, the greater the scouring that occurs.
3. Anempirical equation for maximum scour estimation in front and rear of the structure has been proposed based simple linear regression analysis,using non-dimensional parameter (Sm/ W and $KC\Psi^{0.5}$).

References

1. H.D. Armono, Artificial Reef as Shoreline Protection Structures. *ProsidingSeminar Teori dan Aplikasi Teknologi Kelautan IV*. Surabaya, (2004).
2. B. M. Sumer, et al, Local Scour at Roundhead and Along the Trunk of Low Crested Structures, *Coastal Engineering***52** (2005) 995–1025
3. Carstensen, Stefan dan B. M. Sumer, Scour at Breakwaters Under Combined Waves and Current. *E-proceddings of the 36th IAHR World Congress*, The Hague. Netherlands, (2015).
4. S.A., Hughes, Scour and Scour Protection. *CDCM Training Workshop* Chapter 8. Trinidad, (2001).
5. R. Whitehouse, Scour at Marine Structures: A Manual for Practical Applications. Thomas Telford. UK, (1998).
6. J. Fredsoe and M.B. Sumer, The Mechanics of Scour in the Marine Environment. World Scientific: Singapore, (2002).
7. A.G., Pratikino, Karakteristik Scour di Sekitar Artificial Reef Bentuk Silinder Berongga. Thesis *ITS*. Surabaya, (2015).
8. FvD, Brink. Influence of Liquefaction on Scour around Offshore Monopile Foundations. *Msc Thesis*, University of Twente and Delft University of Technology. Netherlands. (2014).
9. D.M. Young and F.Y. Testik, Onshore Scour Characteristic around Submerged Vertical and Semicircular Breakwaters, *Coastal Engineeering*. **56** (2009) 868–875.
10. Harinaldi. Prinsip-Prinsip Statistik untuk Teknik dan Sains. Erlangga: Jakarta, (2005).

Development of Ecotourism Floating Restaurant in Surabaya City

Eddy Setyo Koenhardono[1,*], *Sardono* Sarwito[1], *RO Saut* Gurning[1], *M. Badrus* Zaman[1], *Alam* Baheramsyah[1], *Indra Ranu* Kusuma[1], *Beni* Cahyono[1], *Juniarko* Prananda[1], *Totok* Yulianto[2], Hasannudin[2]

[1]Department of Marine Engineering, Institut Teknologi Sepuluh Nopember, Indonesia
[2]Department of Naval Architecture, Institut Teknologi Sepuluh Nopember, Indonesia

Abstract. In the 18th century, Kalimas's river serves as a workspace, marketing space, and transportation routes for residents of Surabaya. Although the current condition of Kalimas river has changed a lot, but Kalimas's river has good potential of water tourism and transportation facilities. Therefore, the Surabaya city government setup a policy to revitalize the Kalimas's River. The submarine monument area as one of Surabaya's landmarks next to Kalimas's River will be developed as a 'Riverside Center Business Center'. One feasible way to support the city government program is the development of a floating restaurant. Floating restaurant is just a place to eat, when doing business negotiations or family tours. The concept of floating restaurant development based on ecotourism. Hence, this restaurant is designed environmentally friendly, empowering local people, and become a means of learning and education. The eco-friendly concept is manifested in a flat outward-side asynchronous catamaran form and battery-based electric propulsion systems and solar panels. The menus are provided by local vendors along the river bank of Kalimas so the local food supplier will be more empowered.

1 Introduction

Surabaya's past is a centre for meetings between traders from various corners of Java island with merchants from east and west. At that time, Kalimas's river serves as a workspace, marketing space, and transportation routes for residents of Surabaya [1]. But the role has now been degraded, due to the development of a city focused on land. Whereas Kalimas's river has the potential of water tourism and transportation facilities are very large. Recognizing the hidden potential of the Kalimas River, the Surabaya city government made a policy to revitalize the Kalimas River. The submarine monument or monkalsel area as one of Surabaya's landmarks next to Kalimas River will be developed

* Corresponding author: eddy-koen@its.ac.id

as a 'Riverside Center Business Center'. One feasible way to support the city government program is the development of a floating restaurant. Floating restaurant is just a place to eat, when doing business negotiations or family tours.

The concept of floating restaurant development based on ecotourism. Hence, this restaurant is designed environmentally friendly, empowering local people, and become a means of learning and education. The eco-friendly concept is manifested in a flat outward-side asynchronous catamaran form and battery-based electric propulsion systems and solar panels. This floating restaurant is just a place to eat and meetings, where food is obtained from food vendors along the Kalimas's river, so that food traders are more empowered and create new jobs.

2 Eco-Friendly Floating Restaurants

Floating restaurant development in submarine monument area based on existing floating restaurant design. The characteristics of the floating restaurant design are as follows:

1. Beautiful appearance
2. High level of stability and low rolling motion
3. Low waves generated by the movement of the ship must be as small as possible (low wash displacement craft) in order to minimize the erosion caused by wave wash because the river wall prone to landslides [2].
4. To have the capacity and dimensions that corresponds to the depth of the river, so that food traders are more empowered and create new jobs
5. Silent and low vibration

The characteristics of the floating restaurant is manifested by a catamaran boat that has the form of hull flat side outwards (FSO) with electric propulsion. This form of floating restaurant has several advantages as follows [3-6]:

- a large number of choices of types and shapes with various characteristics;
- Wider deck area.
- more or less high seaworthiness
- required initial stability without restriction of hull aspect ratio
- water-resistant volume above water
- the possibility of slamming the wet deck;
- considerable influence of the external charge across the force
- low wash characteristics boat

Based on the condition of the Kalimas river and requirements the main dimension of the flat side outwards catamaran boat is

- $L_{overall}$: 7,0 m
- $L_{waterline}$: 6,0 m
- $B_{overall}$: 2,8 m
- $B_{demihull}$: 0,79 m
- Depth : 0,5 m
- Draft : 0,29 m
- Displacement : 1,927 ton

And the electric propulsion of the flat side outwards catamaran boat consists of electric outboard motor 500 W, 26 solar panels 80 WP and 10 battery 100 AH [7]. The 3D design of floating restaurant is shown in Figure 1.

Fig. 1. The 3D design of floating restaurant

3 Methodology

The basic material of this environmental friendly floating restaurant is fiberglass. So the analysis of feasibility study of development is based on fiberglass material. The methodology in the feasibility study analysis is divided into three stages, as follows:

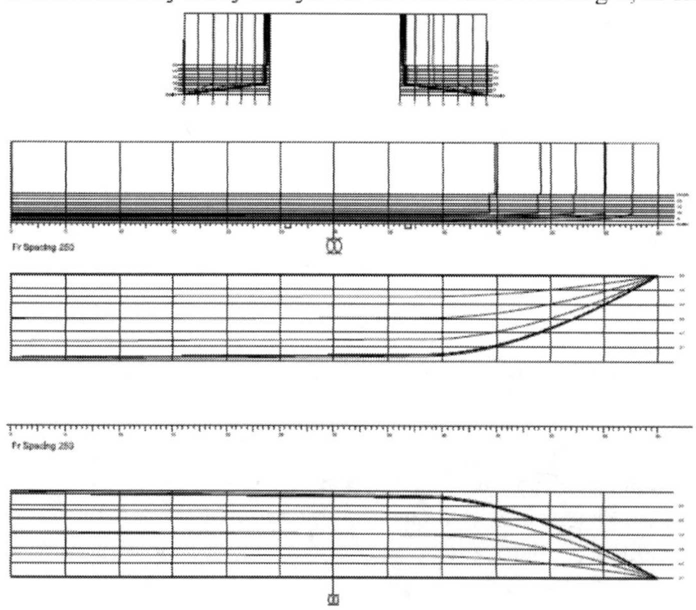

Fig. 2. Lines plan of eco-friendly floating restaurant

3.1 Design of eco-friendly floating restaurant

The design of eco-friendly floating restaurant has a flat side outwards catamaran hull. Flat side outwards catamaran hull has less resistance than monohull or standart catamaran hulls. The other advantage of FSO catamaran has low wash characteristics, therefore the hull is very suitable for floating restaurant in Monkasel area. The lines plan of floating restaurant is shown in Figure 2 and the frame construction can be seen on Figure 3.

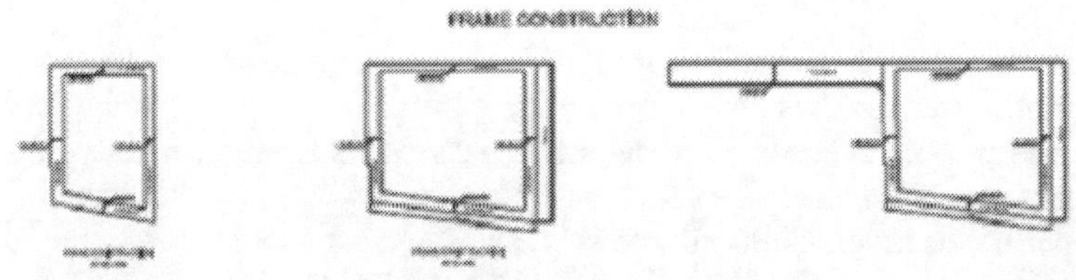

Fig. 3. Construction design of eco-friendly floating restaurant

3.2 Estimation of production cost

Cost components incurred in constructing fiberglass vessels are the cost of hull and construction, miscellaneous equipments, and other expenses. The cost components are listed the Table 1.

Table 1. Hull and Construction

No	Material	Number	Price
	Hull and Construction		
1	Resin	3 drum	16,537,500
2	Catalys	1 kg	60,000
3	CSM 450 g/m2	7 roll	5,460,000
4	WR 600 g/m2	6 roll	5,640,000
5	Cobalt	1 kg	130,000
6	Wax	4 cans	374,000
7	Putty	7 cans	595,000
8	Pigment	2 kg	120,000
9	Foam	14 sheets	4,200,000
10	Mixed gel coat	152 kg	5,700,000
11	PVC Pipe 2 Dim	4 unit	608,000
12	Railing Stainless	3 m	1,200,000
13	Others		3,000,000
	Equipments		
1	Handwash unit	1 unit	250,000
2	Table	1 unit	200,000
3	Passenger chair	10 unit	2,500,000
4	Pilot chair	2 unit	500,000
5	Manual Steering Rotary	1 set	3,500,000
6	Communication Radio VHF	1 set	2,500,000
7	Lamps	1 set	1,000,000
8	Electric Outboard Engine 0,5 kW	1 unit	66,240,000
9	Solar Cell 80 wp	26 unit	30,160,000
10	Battery 200 Ah	5 unit	20,000,000
12	Inverter 1000 Watt	2 unit	16,440,000
13	Mooring line	20 m	590,000
14	Bolder Stainless 3 inchi	1 unit	670,000
15	Life Jacket Standart SOLAS	8 unit	1,400,000
	Miscellaneous		
1	Outfiting & Finishing	1 Package	10,000,000
2	Man hour	5 Package	10,000,000
3	Delivery cost	1 Package	5,000,000
	Total cost		214,574,500

3.3 Estimation of ticketing price

The calculation of the ticket price of each passenger floating restaurant is based on several assumptions as follows:

- Passengers per trip is assumed to be only 5 passengers from a capacity of 8 passengers
- Trips per day are assumed 8 trips
- Hours per trip is assumed 30 minute
- The day of operation in one year is assumed to be 345 days

The initial cost is assumed using 50% owner capital and 50% bank loan, the economic life time of the vessel is 10 years. The maintenance cost considered to be low or close to zero since there is no fuelled engine and the life time of the vessel is only 10 years. The economic feasibility calculation is using NPV concept with several variations of minimum attractive rate of return (MARR) and margin. The margin is profit gained from the difference of return compared to MARR. The saving and loan interest rate are taken from the public bank and the inflation is taken from the data published by the government.

While the assumptions of economic data used are as shown in Table 2.

Table 2 Economic Data

Production cost of the vessel		
Ship price	Rp 214,574,500.00	
Operating costs		
Crew salary	Rp 48,150,000.00	per year
Maintenance and repair	Rp -	per year
Administration cost	Rp -	per year
Insurance	Rp -	per year
Total Operating Cost	Rp 48,150,000.00	per year
Financial Cost / Capital		
Loan term	5	Year
Payment per year	1	times per year
Equity	Rp 107,287,250.00	50%
Debit	Rp 107,287,250.00	50%
Interest Savings / Year	3.75%	per year
Interest Rate / year	10%	per year
Inflation	0.19%	per year
Economic duration of ships	10	years

The ticket price is calculated by using a sofware or Goal Seek and the result is shown in Table 3.

Table 3 Graphic analysis of sensitivity and NPV

MARR	Margin	Ticket price		Inflation	NPV
3.75%	0.00%	Rp 7,370	/ passenger	0.19%	Rp 14,295,720
7.73%	3.98%	Rp 7,625	/ passenger	0.19%	Rp 32,655,015
11.73%	7.98%	Rp 7,907	/ passenger	0.19%	Rp 52,914,741
15.73%	11.98%	Rp 8,213	/ passenger	0.19%	Rp 74,885,320
19.74%	15.99%	Rp 8,540	/ passenger	0.19%	Rp 98,328,835
23.74%	19.99%	Rp 8,884	/ passenger	0.19%	Rp 123,028,982
27.74%	23.99%	Rp 9,243	/ passenger	0.19%	Rp 148,797,648
31.74%	27.99%	Rp 9,614	/ passenger	0.19%	Rp 175,472,050
36.00%	32.25%	Rp 10,021	/ passenger	0.19%	Rp 204,695,947

Based on the variations it can be concluded that the ticket price and NPV are increase along the increase of margin and MARR as can be shown in Figure 4.

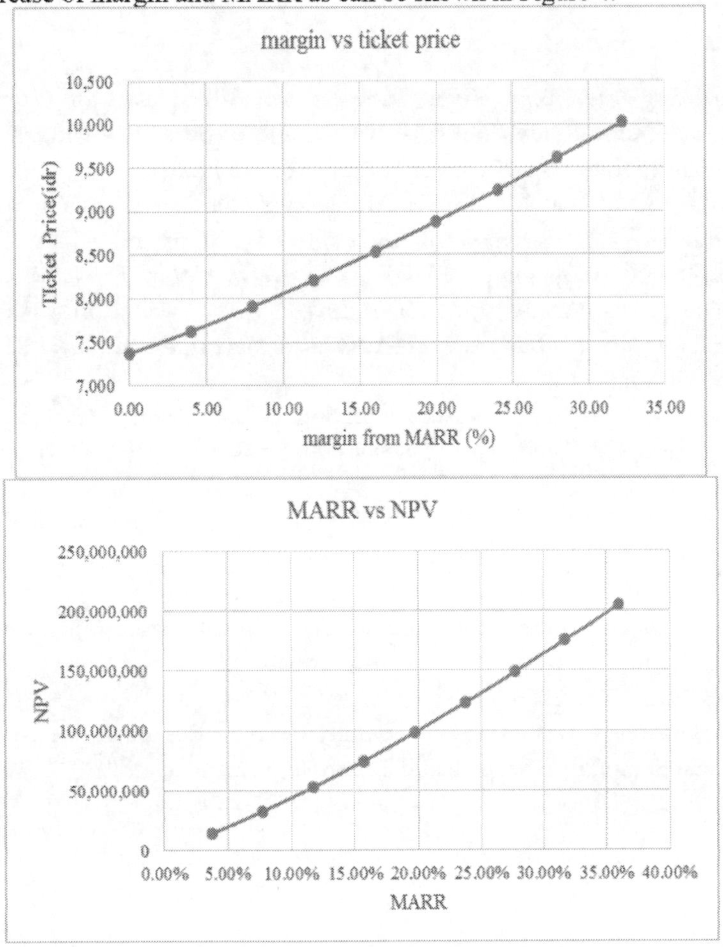

Fig. 4. Ticket Price vs margin (left) and NPV vs MARR (right)

Based on the ticket price data in Table 3, if the ticket price is set at Rp. 10.000,00 per person, obtained

- IRR is 36%, 32.25% higher than the prevailing savings rate
- BEP will be achieved in year 3

Therefore, floating restaurant investment in Kalimas river Surabaya has a very profitable long-term prospect.

4 Conclusion

Based on the feasibility study result, the development of floating restorant in Kalimas's river has high prospect. The ticket price per passenger is only Rp. 10.000,00, so it is very affordable by the people of Surabaya. Where with the low ticket prices, generate IRR 36% and BEP in the third year.

This work was financially supported by PTNBH 2017 of Institut Teknologi Sepuluh Nopember.

References

[1] R. Hastijanti, "Developing River Lane as an Alternative for Water Tourism Destination — Case Study: 'Kalimas' River Lane of Surabaya, Indonesia," in KKU International Engineering, 2014, vol.931, pp. 781–784 (2014).
[2] T. Whittaker, A. Bell, K. M. Morton, M. Shaw, K. Patterson, and M. Surveyor, "An investigation of fast ferry wash in confined waters,"in RINA International Conference on Hydrodynamics of High Speed Craft, 1999, p. 13. (1999).
[3] O. Yaakob, M. Afifi, A. Mukti, A. Nasirudin, "A Low Wash Hullform and Pollutant Free Inland Waterways Leisure Craft,Advan. in Maritime and Naval Scie. and Eng., Constantza, Romania (2010).
[4] O. B. Yaakob, "The wake wash prediction on an asymmetric catamaran hull form,"1st Intern. Conf. on Natural Resources Eng. and Tech., Putrajaya, Malaysia (2006).
[5] O. B. Yaakob, M. Afifi, A. Nasirudin, A. Mukti, A. Arizam, A. Wahab, Hull form configuration study of a low wake wash catamaran leisure boat", 2nd Intern. Conf. on Marine Research and Transportation, Naples, Italy (2007).
[6] O. B. Yaakob, A. Nasirudin, M. P. Abdul Ghani, T. Mat Lazim, M. A. Abd Mukti, M. Ahmed, "Parametric study of a low wakewash inland waterways catamaran,Sci. Iran., 19, 3, Juni (2012).
[7] E. S. Koenhardono, "Design of Eco Friendly Floating Restaurant for River" The 3rd International Seminar on Science and Technology, Surabaya (2017).

Comparative of Feasibility Study Between Diesel Mechanical Propulsion System and Combination of Diesel Engine and Electric Motor Propulsion System on Offshore Patrol Vessel (OPV) 80 m

Eddy Setyo Koenhardono [1,*], and *Amiadji*[1]

[1]Institut Teknologi Sepuluh Nopember, Departemen Teknik Sistem Perkapalan

Abstract. The modern offshore patrol vessel is designed to carry out various missions in the management of economic exclusion zones, thus having some speed of service. For example, the speed for slow patrol is 10 knots, regular patrols 18 knots and chasing 22 knots. This condition requires a flexible propulsion system, which is combination of mechanical and electrical propulsion system. The application of this propulsion system to the OPV can increase the load factor of the propulsion system, resulting in lower fuel consumption than when using a mechanical propulsion system. Based on feasibility analysis results, the choice of combination of mechanical and electrical propulsion system is more advantageous than mechanical propulsion system, although investment and maintenance cost is higher. Therefore, patrol boats should use combination of mechanical and electrical propulsion system.

1 Introduction

Offshore patrol vessels (OPV) have multiple functions, namely military and civilian. Therefore, OPV's weapon system is not as sophisticated as the KCR ship. In the framework of development of Minimum Essential Force (MEF II), the government plans to build 2 OPV vessels of 80 m. The type of OPV vessel propulsion offered by Dutch DAMEN and French DNCS uses a mechanical propulsion system. [1]

The OPV vessel has operational areas in the exclusive economic zone (EEZ) to monitor illegal activities, such as infiltration of foreign troops, illegal fishing, pirate handling, smuggling, and illegal immigrants, handling of marine accidents / oil spills, border controls and other similar activities [2]. Based on this function, the OPV vessel operates under various speed conditions, i.e. surveillance (10 knots), patrols (18 knots), and interception (22 knots).

The CODAE propulsion system is a dynamic combination of diesel mechanic propulsion (DMP) system and diesel electric propulsion (DEP) systems. This system has four propulsion modes: shaft motor, shaft generator, mechanical, and booster modes [3-9]. These various propulsion modes could adapt to meet the requirement of the various OPV 80 m's operation condition. Such propulsion system is worth considering to be applied to the OPV 80 m due the operational flexibility it offers.

This paper presents a configuration layout and specification of the CODAE propulsion system applied as well as the comparison between the DMP and CODAE systems in terms of both technical and economic aspects.

[*] Corresponding author: eddy-koen@its.ac.id

2 The CODAE Propulsion System

Basically, the propulsion system used on ships can be classified into three types, namely mechanical, electrical and combination of them. A propulsion system is a combination of mechanical and electrical propulsion systems, that is called by hybrid propulsion system or combination of diesel mechanic or electric (CODAE) propulsion system. Therefore, the CODAE system can operate as a mechanical, electrical or mechanical and electrical combination system [3-9]. Selection of operating mode on a CODAE propulsion system is adjusted to the existing propulsion power requirement, so that it can maintain to operate at optimal conditions.

A CODAE propulsion system that also uses shaft generator is referred to as a CODAE propulsion system, therefore the system has four operational modes. The four operational modes are electric, power take off/shaft generator, mechanic and hybrid (CODAE) modes. Where the selection of suitable mode is based by fluctuations in the propulsion power and electric power are needed.[4] Figure 1 shows the difference between the mechanical propulsion system and CODAE propulsion system, where the difference lies in the shaft generator / motor and hybrid shaft generator (HSG) drive.

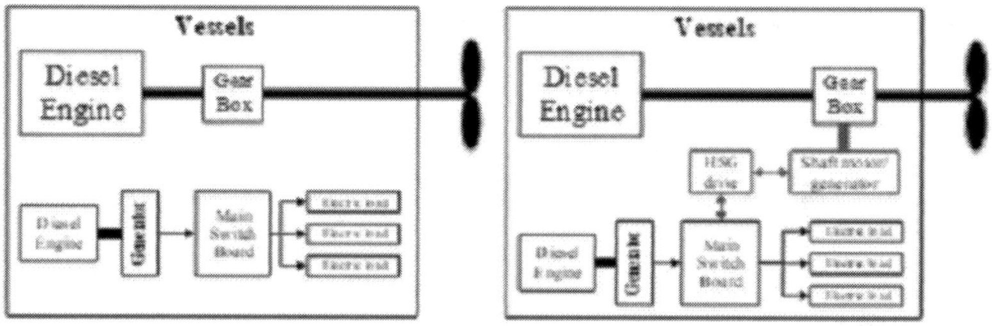

a. Diesel Mechanical Propulsion system b. CODAE propulsion system

Fig 1. DMP and CODAE propulsion system [6]

In general, OPV patrol boats use diesel mechanical propulsion systems, due to their simplicity and reliability. Since OPV ship missions have been increasingly variable, the conventional propulsion system is less suitable. Therefore, the CODAE propulsion system is one of the replacement candidates, in addition to the electrical propulsion system. As mentioned in the above paragraph, the mechanical propulsion system can be a CODAE propulsion system in the presence of a generator / motor shaft and HSG controller.

3 The Feasibility Study of Ship Propulsion System

The feasibility study is a major study to decide whether a project is implemented, suspended or terminated. The review is conducted by reviewing various aspects, such as legality aspect, technical aspect, marketing, social economy, economics and management and finance. In this paper, the comparison of feasibility analysis between the use of mechanical propulsion system and CODAE on the OPV patrol boat is only viewed from the technical and economic aspect. The purpose of the technical aspect is determination of propulsion and electric system configuration for both DMP and CODAE for the OPV 80 m. The economic aspect includes comparing total investment cost, fuel consumption cost, and maintenance cost between the use of both propulsion systems. The result of comparison is a break even point (break event point).

Tables 1. Operational Profile of OPV 80 m

No.	Operation conditions	Vs [knot]	Propulsion Power (kW)	Electrical Power (kW)	Duration [hour/year]
1	*At port*	-	0	873	3504
2	Loitering	10	645	1021	1577
3	Patrol	18	3921	1110	2102
4	Interception	22	7637	1129	1577

3.1 Technical Aspect

The configuration of OPV's propulsion system is obtained based on OPVs' operational profile. The OPV's operation profile is as shown in Table 1, there are four operation conditions with different speed. In accordance with the need for propulsion power and electrical power under each operating condition of the vessel, the engine propeller matching (EPM) analysis, and generator capacity analysis are performed. There are three main purposes of EPM analysis, first, the full power of the main engine can be developed or almost full power. Secondly, the propulsion plant will function satisfactorily in all designs and design conditions. Third, the main engine operates on the best fuel consumption. The result of both analyses for DMP and CODAE propulsion system is presented in Table 2.

Tables 2. Component of DM and CODAE Propulsion System

No	Component	DM Propulsion Systems		CODAE Propulsion System	
1	Main Engine	2	MTU 20V4000 M93L, 4300kW, 2100 rpm	2	MTU 20V4000 M73L, 3600kW, 2050 rpm
2	Auxiliary Engine	4	CAT C18 ACERT, 450 kWe, 1500 rpm, 380VAC.	4	KOHLER KD800-YF, 720 kWe/900 Kva, 1500 rpm, 400VAC.
3	Propeller	2	Wageningen B4.65, D=1,83 m, P/D 0.852, η_0=0.569	2	CPP B 4.55, Ø 1,6 m, *P/D* 0,5-1,4
4	Gear box	2	Single input & output, ratio 3:1	2	ZF 24060 D PTI, 4615 kW/2050 rpm, single input & double output (1:1 & 2.577:1). 2650 kg
5	Shaft motor/ generator			2	ABB M3BP 355 SMC4, 400 VAC, 616 A, 355 kW, 50 Hz, 1487 rpm, 2280 Nm
6	HSG drive			2	ABB ACS850-04-650A-5, 3 phase, 355 kW, 380~500 VAC, 650 A

3.2 Economic Aspect

3.2.1 Investment Cost

The investment cost of a propulsion ship system OPV 60 m is the cost that must be incurred by the ship owner to build or modify the ship's propulsion system. This cost includes the components of the configuration of each type of propulsion system. The components of the mechanical propulsion system and CODAE refer to Table 2, which consists of main engine, auxiliary engine, propeller, gear box, generator shaft / motor and HSG drive.

Generally, the investment cost of each component of the ship propulsion system is based on the power capacity. For investment cost of diesel engine as main driver of ship besides influenced by big power factor, hence congress type conguration and bore size also have an effect. For example, the investment cost of an inline diesel engine is higher than that of a V-line diesel engine. In addition, the investment cost of V-line diesel engines decreased along with the increase in bore size. Table 3 summarizes the prices of ship propulsion system components.

Tables 3. Comparasion of investment cost between DMP and CODAE Propulsion System

No	Component	DMP Propulsion Systems	CODAE Propulsion System
1	Main Engine	€ 2.924.000,-	€ 2.448.000,-
2	Auxiliary Engine	€ 720.000,-	€ 1.152.000,-
3	Propeller	€ 860.000,-	€ 791.000,-
4	Gear box	€ 598.000,-	€ 609.500,-
5	Shaft motor/generator		€ 95.850,-
6	HSG drive		€ 56.000,-
Total of investment cost		€ 5.102.000,-	€ 5.133.720,-

3.2 Fuel Consumption Cost

In general, the largest component of ship operating costs is fuel costs (Wijayanto, 2011). One of the factors affecting the fuel cost of a ship is load factor of main engine and auxiliary engine. Operation of main engines and auxiliary engines at high load factor conditions have lower SFOC than operating at low load factor. OPV vessels operate at some service speed according to their mission, so the load factor varies. The comparison of fuel consumption between the application of mechanical propulsion system and CODAE according to the operational profile of OPV 80 m in Table 4.

Tables 4. Fuel consumption based on OPV profile

Operation mission	Speed (knot)	Duration (hour/year)	DMP propulsion system		CODAE propulsion system	
			Main engine	Aux engine	Main engine	Aux engine
Harbouring	-	3500	-	998 ton	-	645 ton
Surveillance	10	1550	288 ton	445 ton	-	571 ton
Patroling	18	2100	1836 ton	593 ton	2040 ton	245 ton
Chasing	22	1550	2212 ton	445 ton	2204 ton	605 ton
Total of fuel consumption			6817 ton/year		6309 ton/year	

3.3 Maintenance Cost

The cost of maintaining a patrol boat propulsion system includes all costs incurred for maintenance and repair of the propulsion system, including maintenance costs of the main engine, diesel generator set, CPP, gearbox, frequency converter, and generator / motor shaft. Maintenance cost of each component is primary data obtained through direct interview with the expert from the engine distribution's company, shipyard, and shipping company as owner and / or ship operator. The result of interview may be tabulated in Table 5 and Table 6.

Table 5. Maintenance and spare part replacement cost of diesel motor

Diesel motor	Type of maintenance	Maintenance cost	Replacement cost of spare part
Main engine: 3440 kW, 2100 rpm	Top Overhaul	Rp 960.000.000,-	Rp 3.602.000.000,-
	General Overhaul	Rp 1.601.000.000,-	Rp 6.003.000.000,-
Diesel generator: 565 kW, 1500 rpm	Top Overhaul	Rp 143.000.000,-	Rp 792.000.000,-
	General Overhaul	Rp 187.000.000,-	Rp 1.322.000.000,-
Diesel generator: 846 kW, 1500 rpm	Top Overhaul	Rp 175.000.000,-	Rp 1.189.000.000,-
	General Overhaul	Rp 292.000.000,-	Rp 1.983.000.000,-

Tabel 6. Maintenance cost of electric motor based on spesification and the level

Spesification of electric motor	Level	Hour operating	Recondition cost
400 kW; 400 Volt; 1500 rpm; 50 Hz;	L1	10.000	Rp 6.500.000,-
	L2	20.000	Rp 6.500.000,-
	L3	40.000	Rp 36.000.000,-
	L4	80.000	Rp 68.000.000,-
700 kWe; 400 Volt; 1500 rpm; 50 Hz;	L1	10.000	Rp 9.500.000,-
	L2	20.000	Rp 9.500.000,-
	L3	40.000	Rp 57.000.000,-
	L4	80.000	Rp 89.000.000,-

The amount of labor honor for the maintenance of one unit of perporosan system based on field data in a shipyard in Surabaya is Rp. 23.343.750, -. In general, shaft system maintenance should replace three packings, where the price per unit is Rp 2.500.000, -. This replacement is done when the ship does General Overhaul 5 years.

4 Break Event Point

Break Even Point (BEP) between mechanical and hybrid propulsion systems is a point where the sum of investment costs and operational costs are both the same. The length of time to reach Break Even Point can be used to make changes to the ship propulsion system or not. Break Even Point calculations require basic calculation components such as the following:

1. Fixed Cost. In the ship propulsion system investment, this component is a fixed or constant cost.
2. Variable Cost. In the ship propulsion system investment, the dynamic cost components depend on the ship's operational profile. If the ship's operational length and speed increase, then the variable cost increases. Examples of these costs are the cost of fuel consumption, maintenance costs and replacement cost of spare parts.
3. Selling Price. This component in the ship propulsion system changes is an advantage that can be obtained due to changes in the propulsion system.

In this analysis, there are only two component of BEP to be considered, that is fixed and variable costs. The fixed cost is investment cost of the propulsion system. The investment cost includes purchasing and installation costs of components of both the propulsion systems. Variable cost includes fuel consumption costs dan all maintenance cost , both costs are time dependent. Figure 2 is the result of BEP analysis, where the BEP can be reach less than one year.

Conclusions

As shown in Table 2, the main differences between DMP and CODAE for OPV's 80 are reductions of main engine power and increase the diesel generator power capacity. The use of electric motors as the propulsion of ships under surveillance conditions can reduce the fuel consumption is very large. Operation of diesel motors on low load factors (15%) results in high fuel consumption (300 gr/kWh). The difference in investment cost between DMP and CODAE propulsion system is not much different, compared to when using electric propulsion system. Therefore, the BEP between using a mechanical propulsion system into a CODAE propulsion system can be achieved in less than a year. Therefore, in the future, the use of CODAE propulsion systems on patrol boats is highly recommended.

This work was financially supported by PTNBH 2016 of Institut Teknologi Sepuluh Nopember.

References

[1] A. D. Firmansyah, et al.,"Perancangan*controllable pitch propeller* pada kapal *offshore patrol vessel* 80 (OPV80)," Jurnal Teknik ITS, vol. 1, no. 1, pp. G230 – G231, Sept. 2012.

[2] M. J. Underwood, "Patrol boats: aworldmarket overview,"Marine Technology, vol. 33, no. 2, pp. 101 – 107, Apr. 1996

[3] Koenhardono, E.S., "Neural Network-Based Engine Propeller Matching (NN-EPM) for Trimaran Patrol Ship", in Applied Mechanics and Materials Vol. 493 (2014) pp 388-394.

[4] D. Wahyudi, "Hybrid Propulsion Sistem (DMP & DEP) for Trimaran Type Fast Patrol Boat", SENTA Conference, Surabaya 2009.

[5] Koenhardono, E.S., "Perencanaan Konfigurasi Sistem Propulsi Hybrid Pada Kapal Patroli Trimaran", SENTA Conference, Surabaya 2012.

[6] B. Kwasieckyj,"Hybrid propulsion systems: efficiency analysis and design methodology of hybrid propulsion systems," TU Delft, SDPO.13.008.m, March 2013.

[7] Koenhardono, E.S., "Selection of propulsion systems for naval patrol trimaran vessels between the hybrid shaft generator and mechanical propulsion system", SENTA International Conferenece, Surabaya, 2016.

[8] Koenhardono, E.S., "Applications of Hybrid Propulsion System on Trimaran Tourist Vessels that Reliable and Environmental Friendly", SENTA International Conference, Surabaya 2017.

[9] Koenhardono, E.S., "The Study of the Application of Hybrid Propulsion System on OPV with Controllable Pitch Propellers", IJMEIR, Vol. 1(4), Sept. 2017. 346-354.

MATEC Web of Conferences **177**, 01012 (2018) https://doi.org/10.1051/matecconf/201817701012
ISOCEEN 2017

Analysis of the buckling moment on rectangular hollow pipe under pure bending load

Hartono Yudo[1,*] , *Wilma* Amiruddin[1], *Sarjito* Jokosisworo[1]

[1]Naval Architecture Departement, Diponegoro University, Indonesia

Abstract. Not only the circle hollow pipe but also the rectangular hollow pipes are used on offshore structure. In this study, the bending moment was given at both end of pipe. The strength of buckling moment can be reduced by increasing the length. The rectangular pipe models are varying from a/b = 1, 2, 4; a/t = 10, 15, 20 and L/a = 10, 15, 20. On the rectangular pipe which is have larger rectangle area (a/b), have the larger deformation. Vice versa on the rectangular pipe which is have smaller rectangle area (a/b), as though on a/b = 2 and a/b = 4 which is have smaller deformation. The conclude that is the larger one of rectangle area then the oval deformation is large too. The buckling strength will increase with decreasing a/b and deformation at mid span will be increase. The buckling strength decrease with increasing of L/a and a/t. The strength of the buckling moment in the elastic state will decrease with increasing of a/t and a/b. For a/b = 1, the strength of buckling moment will tend to be equal at L/a = 10-20. Pipe with increasing of a/t, L/a and a/b will be elastic.

1 Introduction

The damage of material when buckled is difficult to be discovered, because that is suddenly occurred. The steel is one of the important material constructions. Particularly have mechanical properties which are strong enough and the ductility. Ductility is the material ability to deform, withstand the tensile and compress before failure.

The steel have random circumstances when used. If the maximum ability of construction has been discovered, thus the construction is safe. Buckle occurred due to pressure on the steel bar which has deformation in this case through an axial compressive force. Buckle can be occurred before or after maximum stress was reached. It would not a problem if the buckle occurred after the steel passed the maximum stress. However if the buckle occurred before the steel passed maximum stress, it can be a problem because the buckle will suddenly occurred.

The structure with the rectangular hollow pipe has been used on house roof construction, the bridge pillar, port construction, the offshore construction and building construction. The typical of the neither offshore construction nor coastal construction is the casing for the pile; the offshores leg, and the funnel. The advantage of rectangular hollow pipe is the behaviour which is buckled from two directions and particularly strong enough to withstand the twisting.

* Corresponding author : hartono.yudo@yahoo.com

In the followings, the formulas for estimating buckling stress and moment of rectangular hollow pipe subjected to bending are briefly explained.

The maximum bending stress of a rectangular pipe under the critical (buckling) moment Mcr can be expressed by Eq. 1:

$$\sigma_{cr} = \frac{M_{cr} C}{I} \tag{1}$$

where I is inertia moment and C is distance from neutral axis.

Timoshenko and Gere [1] expressed that the maximum compressive stress at the critical buckling moment is about 30 % higher than that obtained from Eq. 1.

The previous investigation by authors (Yudo and Yoshikawa [2]) for a straight pipe the critical bending moment in linear calculation is about 10 % higher than that obtained from Eq. 1.

The initial yield moment of a pipe under bending is shown below:

$$M_Y = \sigma_Y Z \tag{2}$$

Where σ_Y is the yield stress and Z is section of modulus.

Hauch and Bai [3] has earned a set of equations for calculating the maximum allowable bending moment including proposed safety factors for different targeted safety levels.

The investigation by authors (Yudo and Yoshikawa [4]) obtained the calculation results, the maximum moment for a long pipe considering oval deformation is shown as follows:

$$M_{max} = 052\ M_{cr} \tag{3}$$

Yudo and Yoshikawa [5] has already investigated the imperfect pipe subjected to bending, the reduction of buckling strength due to imperfection is little bit greater in the case of large D/t than of smaller D/t.

Essentially the rectangular hollow pipe is not only encountered the axial load. If the truss bar loaded rapidly with axial tensile force, thus the truss bar will deform. The deformation from straight circumstance of bar to be curved circumstance that told buckling.

From the mechanic of materials known that only the very short rectangular hollow pipe can be loaded until reached the yield strength, whereas the general circumstance, the spontaneous bending due to the unstability occurred before the strength of truss bar has reached that circumstance which is told buckling.

To determine the strength of rectangular pipe, the circumstance of rectangular pipe must be gave the material

Buckling can be defined as a failure phenomenon which is occurred due to compressive load with the result that causing the deformation which have the shape of lateral deflection to another equilibrium form.

Buckling phenomenon can be divided to be two sections, the global bending and local bending. An example of global bending is the entire structures have curved as one unit, whereas the local bending is the bending which is occurred at the plate elements.

To obtained critical load value, the eigenvalue should be input to this simple equation:

$$P_{cr} = P_{applied} \times \text{Eigenvalue} \quad (4)$$

Therefore, the eigenvalue which is generated also can be interpreted as the factor of safety which is having by the structure. Decreasingly of the eigenvalue which is generated, thus greater of the failure probability due to buckling. Therefore, the greater of eigenvalue expected on the analysis of structural strength to show the factor of safety.

2 Calculation result

The typical buckling mode, which is obtained from nonlinear calculation, is shown in Fig. 2. The stress at the lower side of the cylinder is compression, and that at the upper side is tension. The oval deformation is larger in the vicinity of midspan than at the span end. Therefore, the region of buckling is limited in the bottom area close to the center.

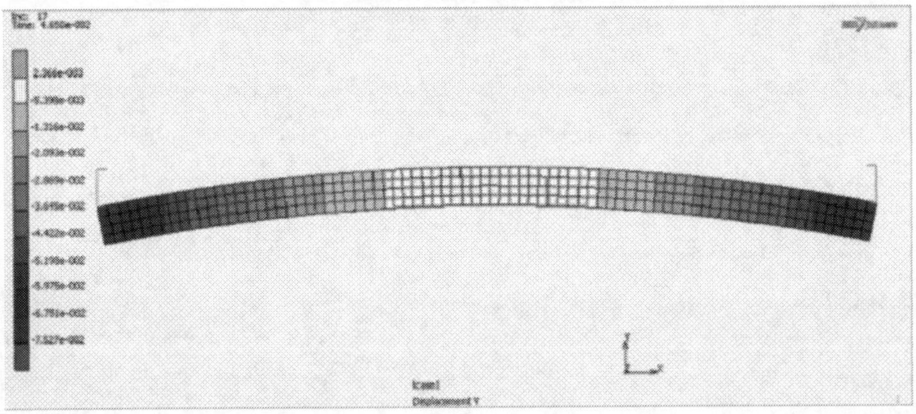

Fig.1. Rectangular pipe deformation

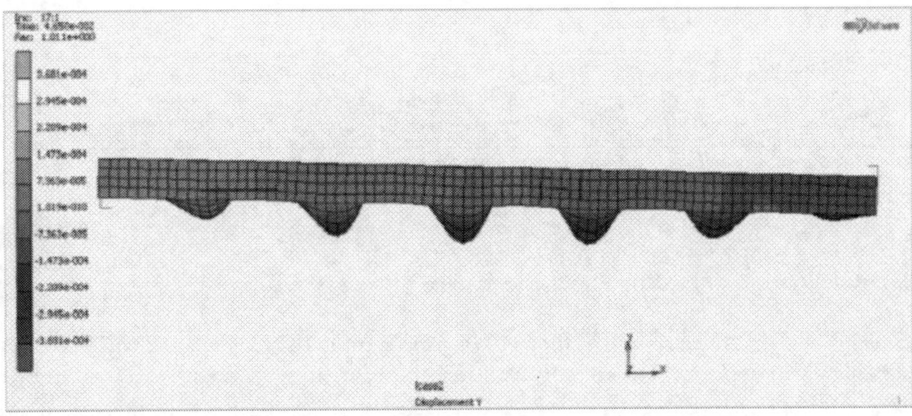

Fig.2. Rectangular pipe buckling

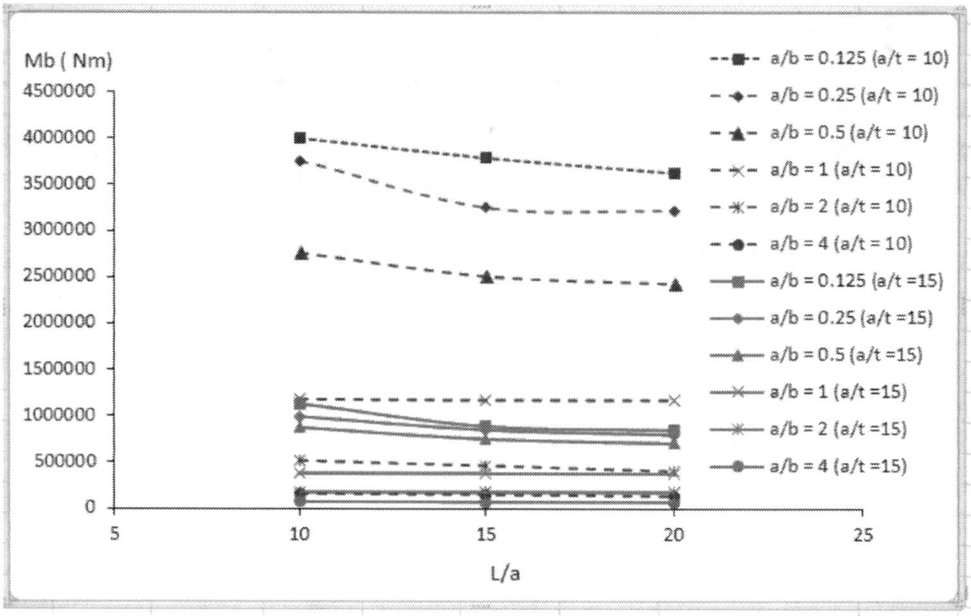

Fig.3. Relationship of buckling moment Holow pipe and L/a

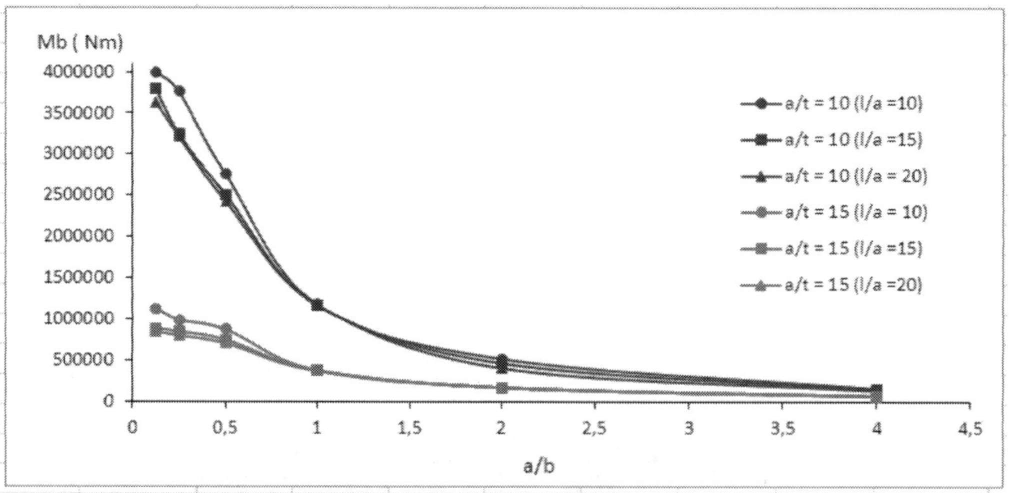

Fig.4. Relationship of buckling moments Holow pipes and a/b

Figure 4 shows that the strength of the buckling moment in the elastic state will decrease with increasing of a/t and a/b. For a/b = 1, the strength of buckling moment will tend to be equal at L/a = 10-20.

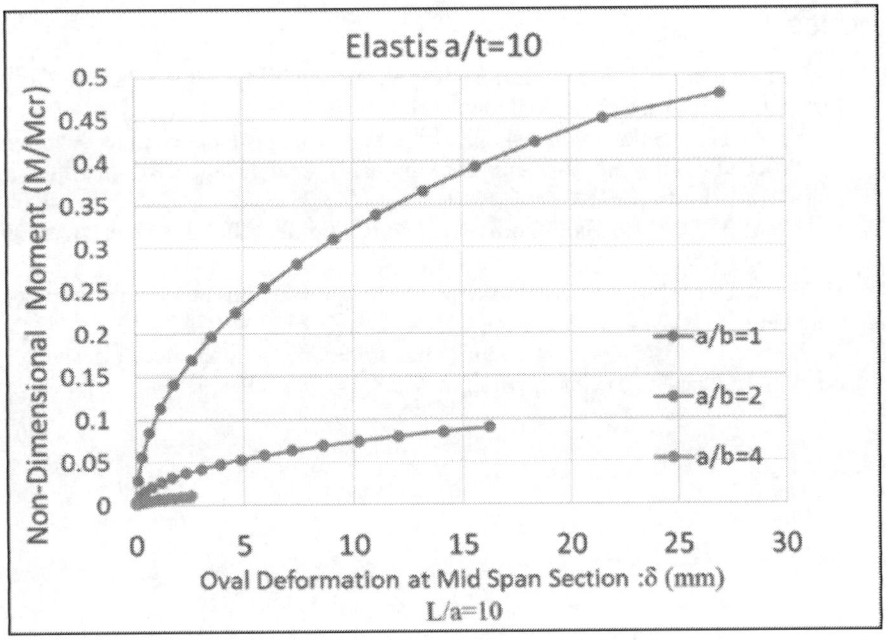

Fig.5. The deformation on middle of rectangular pipe (elastic)

On the rectangular pipe which is have larger rectangle area (a/b), have the larger deformation. Vice versa on the rectangular pipe which is have smaller rectangle area (a/b), as though on a/b = 2 model and a/b = 4 model which is have smaller deformation. Therefore, the larger one of rectangle area the oval deformation is large too. Figure 6 shows the difference of deformation on mid-span for each rectangular pipe.

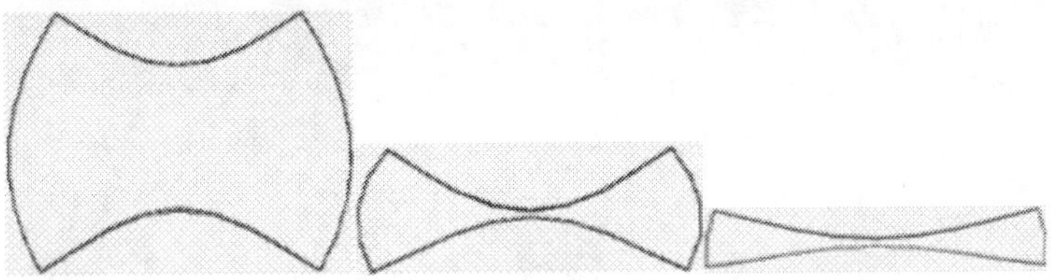

Fig.6. The deformation on mid-span of rectangular pipe (elastic)

Conclusion

In this study, the series calculations of buckling and collapse strength of rectangular hollow pipe under bending are performed by utilizing nonlinear FE software. The typical buckling modes for rectangular hollow pipes are obtained by numerical calculations. The followings are clarified by the numerical calculations.

1. The region of buckling is limited at the compressive side in the vicinity at mid span
2. The strength of the buckling moment in the elastic state will decrease with increasing of a/t and a/b. In Figure 4 shown for a/b = 1, the strength of buckling moment will tend to be equal at L/a = 10-20.

Reference

1. S. Timoshenko, J. Gere., Theory of elastic stability, 2nd edn. McGraw-Hill International Book Company, New York (1961).
2. H. Yudo, T. Yoshikawa., Mechanical behaviour of pipe under pure bending load. In: Proceedings of the 26th Asian technical exchange and advisory meeting on marine structures, pp 359–364 (2012).
3. S. Hauch, Y. Bai., Bending moment capacity of pipe. Offshore Mech Arct Eng PL-99-5033 (1999).
4. H. Yudo, T. Yoshikawa., Buckling phenomenon for straight and curved pipe under pure bending. J Mar Sci Technol, vol 20, issue1, pp 94-103 (2015).
5. H. Yudo, T. Yoshikawa., Buckling phenomenon for imperfect pipe under pure bending . J Mar Sci Technol, vol 20, issue 4, pp 703-710 (2015).

System Impressed Current Anti Fouling (ICAF) Against Micro Fouling (Bacteria) on Ship's Cooling System

Herman Pratikno[1,*], *Harmin* Sulistiyaning Titah[2] , and *Muhammad* Danesto Rizky Mauludin[1]

[1]Departement of Ocean Engineering, Faculty of Marine Technology, Institut Teknologi Sepuluh Nopember, Keputih, Sukolilo, 60111 Surabaya, Indonesia
[2]Departement of Environmental Engineering, Faculty of Civil, Environmental and Geo Engineering, Institut Teknologi Sepuluh Nopember, Keputih, Sukolilo, 60111 Surabaya, Indonesia

Abstract. Bio-fouling is the attachment and accumulation of an organism or micro-organism to a material. Bio-fouling is one of the causes of declining performance and quality in a system of structures, especially structures that have a direct relationship with the waters where the bio-fouling live and move. The research was conducting using a simple circuit that represents Impressed Current Anti Fouling (ICAF) system which is generally located on the cooling of the ship against micro-fouling or bacteria. Species of bacteria (*Pseudomonas fluorescens*) in the marine environment was used as an object. The bacteria were introduced into the reactor containing the cathode and anode of this simple system, with inputs of experimental time variation of 3, 5 and 7 minutes and electrical current variation of 0.1, 0.3, and 0.5 Ampere. The determination of bacteria population was conducted using standard plate count methods. The results showed that ICAF system can reduce the bacteria populations. The largest percentage of *Pseudomonas fluorescens* reduction was 99.9%, while the smallest percentage showed 98.5% percentage. In conclusion, the simple of ICAF system can be used to prevent appearance of micro-fouling in marine environment.

1 Introduction

The most commonly material in the offshore technology is definitely from metal. Metals have different corrosion resistance. The problem with the oxidation of a metal material lies in almost every element on earth, such as water or air. Corrosion caused by water, especially seawater on an offshore structure is inevitable. The sea water has a rapid corrosion rate, caused by salinity present in seawater or solid and gaseous solutes, such as inorganic salts, dissolved gases and organic compounds of living organisms. Sea water can dissolve other substances in large quantities than other liquids. Therefore, the anti-corrosion system is needed by all structures which directly contact to sea water.

* Corresponding author: hermankelautan@gmail.com

Corrosion due to algae and bacteria is the most common in the category of biological corrosion. The affects of microorganisms on biocorrosion in the marine environment is more than macroorganisms. The bio-corrosion or microbial influenced corrosion (MIC) occurs due to the fixation of bacteria, release of metabolites and formation of biofilms that induce or accelerate the corrosion process [1].

Bio-fouling that was caused by microorganism or it was called as biofilm is ubiquitous in aquatic environments. Micro-fouling were accumulation of many types of microorganisms such as bacteria, fungi, algae, diatom, protozoa etc.,. Bacteria is the first and predominant type of microorganisms in the marine environment [2]. According to Donlan [3], biofilm is an accumulation of microorganisms that is irreversibly associated with a surface and enclosed in a matrix of primarily polysaccharide material. A single bacterium can form biofilm, but in natural environment biofilms are formed by a wide group of organisms. Moreover, it is becoming clear that these natural assemblages of bacteria within the biofilm matrix function as a cooperative consortium [4].

Many microorganisms can produce polymeric substances outside the body (exudates); these are called as extracellular polymeric substances (EPS). EPS is called by different names, such as slime, mucus, mucilage, glycocalyx and exopolymer [5]. The major component of the EPS is sugars (including neutral sugars, uronic acids and hexosamines).*Pseudomonas* is one of genus that can produce EPS [6]. *Pseudomonas* (and other aerobic bacteria) have been shown to both increase and decrease the rate of metal corrosion through polymer production during biofilm formation [7,8]. *Pseudomonas* sp. 200 was known to accelerate corrosion with increasing growth rate [9]. *Pseudomonas* sp that isolated from products of corrosion of diesel pipelines and metal surfaces immersed in fresh water and sea water could produce EPS that favors the formation of biofilm [1]. *Pseudomonas fluorescens* as an example of bacteria in the marine environment of shallow depth (pelagic) which of course as the location where the hull of the ship is located. *Pseudomonas fluorescens* is one of species of bacteria that can cause the microfouling. *Pseudomonas fluorescens* SBW25 produced biofilms that were comprised of extensive, extracellular non-periodic webs of fine (< 20 nm wide) cellulose fibers [10]. According to Baum et al. [11], polysaccharides were found to make up a smaller fraction of the biofilm mass (ca. 15% wt), and were of the magnitude expected in a vegetative bacterial cell. These results are atypical for EPS produced by Pseudomonads, which generally have a higher sugar-protein ratio. However, based on Raza et al. [12], the EPS that produced by a newly isolated strain *Pseudomonas fluorescens* WR-1 was composed of arabinose, glucose and uronic acid.

Based on our previous study, 3 species of bacteria *Escherichia coli, Pseudomonas fluorescens,* and *Thiobacillus ferroxidans* can caused bio-corrosion on steel structures of ASTM A106 and A53 in deep seawater (salinity of 33‰), medium seawater (salinity of 35‰), and shallow seawater (salinity of 37‰) [13] The bio-corrosion rate by *P. fluorescens*on Aluminium Alloy 6063 at salinity of 37‰ increased by one point six-fold compared with the condition without bacteria addition at the same salinity [14].

Sacrificial Anode, Impressed Current Cathodic Protection (ICCP) and Impressed Current Anti Fouling (ICAF) are three ways to avoid or slow down corrosion on a metal. The use of anti-corrosion systems such as ICCP and ICAF should continue to be developed so that a shortage of sacrificial anode system can be decreased. New innovation in it is also urgently needed to improve the effectiveness and efficiency of the corrosion prevention tool. ICAF is short of impressed current anti-fouling. Its function is to prevent or inhibit the growth of fouling, which in this case is caused by marine biota, such as barnacles, shellfish, algae, and other marine plant species. Fouling is very detrimental because it attaches to the hull of the ship so that it can increase the prisoner of the ship. Research on how effective the system is killing a fouling is still limited. The aim of the research was to determine the decreasing precentage of population of *Pseudomonas fluorescens* using simple ICAF system.

2 Materials and methods

2.1 Preparation of ICAF System

A simple ICAF System was designed based on adoption from ships cooling system. In this system, the DC current was be used due to the supply the anode and cathode must be distinguished between positive and negative current sources. Both the anode and the cathode are various types, depending on the material to be protected and how long it will be used. The metal of Cuprum (Cu) was be used as anode with dimension of 15 x 15 x 1 cm and he material of high-strength low-alloy (HSLA) steel was be used as cathode. Figure 1 showed the ICAF system.

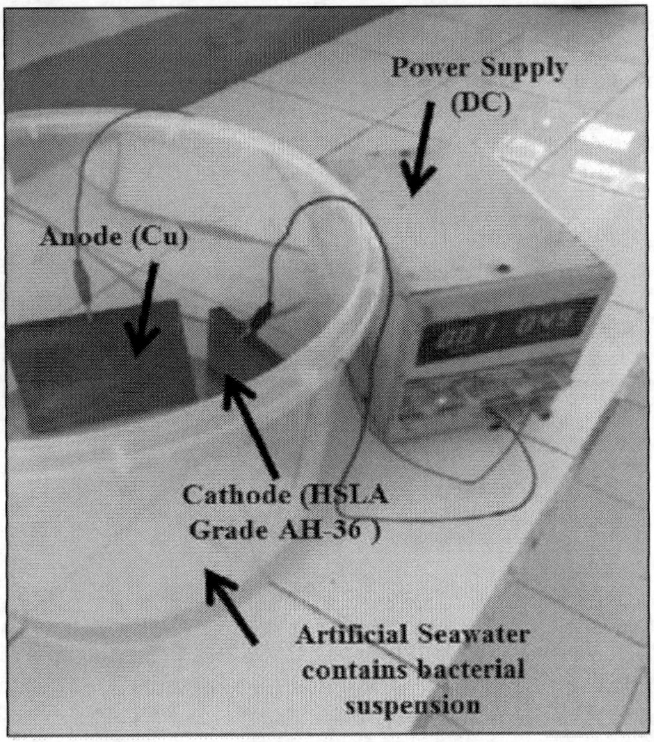

Fig. 1. Simple of ICAF System.

2.2 Preparation of artificial seawater

This research used a chemical solution instead of sea water with a salinity of 35 ‰. Salinity of 35‰ is the salinity in the ocean of medium depths, where in the microorganism commonly found at this depth, including bacteria. The chemical composition of seawater replacement is in accordance with ASTM D1141-90, 1994 [15].

2.3 Preparation of Bacteria

The preparation of bacteria was conducted based on Pratikno and Titah [16]. The pure culture of *Pseudomonas fluorescens* was be inoculated onto nutrient agar (NA) media using streak plate technique based on Harley and Prescott [17]. The age of bacteria for the test was 24 h. After that, one colony of bacteria was transferred to nutrient borth (NB) and keep in shaker incubator of Innova 2000 (New Brunswick-Eppendorf, Germany) at 150 rpm and room temperature, 33 °C for 24 h. The cell suspension of selected bacteria was prepared by harvesting the cells at the middle of the logarithmic phase, based on the typical of growth rate graph for the selected bacteria. At this time, the OD at 600 nm was 1.0 was determined using UV spectrophotometer Genesys 20 (Thermo, USA). The cells were harvested through centrifugation of Jouan E82 (Thermo, USA) at 4,000 rpm for 15 min. The obtained pellet was then washed twice using 8.5 g NaCl/1000 mL solution.The suspension of bacteria was ready to be used in ICAF system test.

2.4 Determination of Bacterial Population

Colony-forming units (CFU or cfu) is a measure of viable bacterial. CFU measures only viable cells by direct microscopic counts (cell counting using bacterial colony counter or BBC). For convenience, the results are given as CFU/mL (colony-forming units per milliliter) for liquids, and CFU/g (colony-forming units per gram) for solids. The CFU was conducted using pour-plate technique based on Harley and Prescott [17]. Calculation of the number of bacteria (CFU) per milliliter or gram of sample by dividing the number of colonies by the dilution factor. A wide series of dilutions (e.g., 10^{-4} to 10^{-10}) is normally plated because the exact number of live bacteria in the sample is usually unknown. Greater precision is achieved by plating duplicates or triplicates of each dilution. That is, the final plates in the series should have between 25 and 250 colonies. Fewer than 25 colonies are not acceptable for statistical reasons, and more than 250 colonies on a plate are likely to produce colonies too close to each other to be distinguished as distinct CFU. The number of colonies per ml reported should reflect the precision of the method and should not include more than two significant figures. The CFU/ml can be calculated using the formula:

$$\text{CFU/ml} = \frac{\text{no. of colonies x dilution factor}}{\text{volume of culture plate}} \quad \text{.......} \quad (1)$$

2.5 Running of ICAF System

Running test of ICAF system were carried out based on the time variation (3, 5 and 7 min) and electrical current (0.1, 0.3, and 0.5 Ampere). The main parameter was the population of *Pseudomonas fluorescens* at initial time and after operation of ICAF system. The determination of population of *Pseudomonas fluorescens* was conducted using pour plate count methode based on Harley and Prescott's standard plate count [17].

3 Results and Discussion

Figure 2 showed the results of simple ICAF system running. Based on this graph, reduction of *Pseudomonas fluorescens* population was higher during the time increased. Hong et al. [18] reported that fluorescently-tagged *Pseudomonas aeruginosa* bacterial cells were adhered onto a glass slide with a conductive layer of indium tin oxide (ITO). Anodic and cathodic currents of 15 uA/cm2 were be applied to the ITO layer, and bacterial detachment was monitored with a fluorescent microscope. After 20 minutes of application of the cathodic current, the total number of attached bacteria dropped to 20%. At the 40-

minute, the population on the surface with anodic current fell to approximately 70%. Based on Gabi et al. [19], anodic microcurrents with current densities of 75 nA/mm2 and 750 nA/mm2 were shown to increase the formation of biofilms on the surface of a platinum (Pt)- coated quartz crystal.

The decreasing of *Pseudomonas fluorescens* population trends were occurred at all variation of electrical current (0.1, 0.3, and 0.5 Ampere). According to Berthelot [20], the effect of current applied across a substrate's surface has also been shown to affect bacterial attachment. The electrical current of 0.07 and 0.125 mA reduced the motility of *Pseudomonas aeruginosa* and *Escherichia coli,* indicating that there were be an optimal range of current that will predictably and consistently impact to bacterial species.

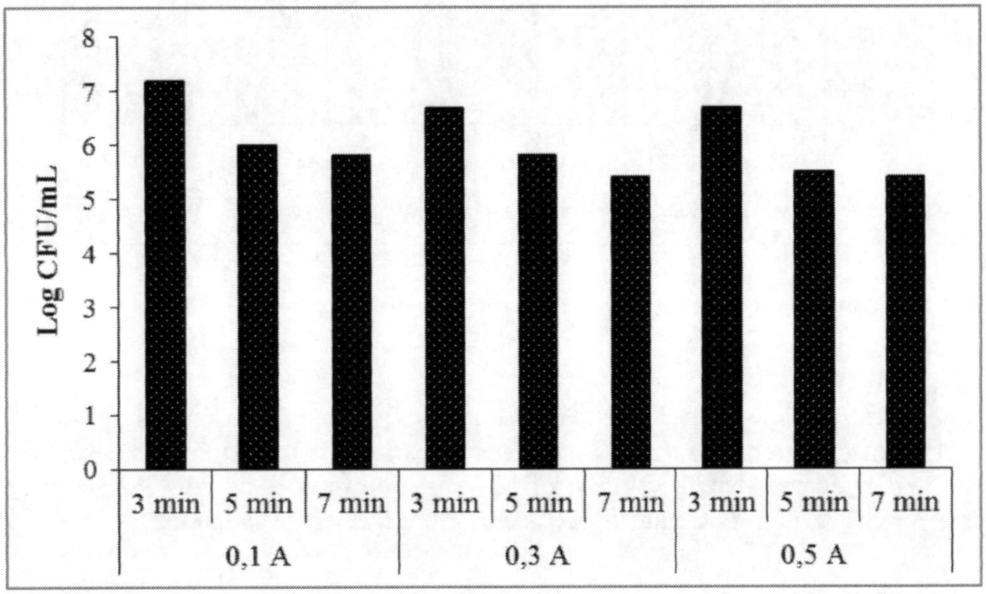

Fig. 2. Population of *Pseudomonas fluorescens* population at variation of electrical current after 3, 5, 7 min in ICAF system.

Based on Figure 3, the lowest percentage of *Pseudomonas fluorescens* population reduction occured at combination of variables with duration time of 3 minutes and electrical current of 0.1 (98.5%), while the largest percentage reached 99.9% at combination of variables with duration time of 7 minutes and electrical current of 0.5 A. The reason this number is not zero is because although bacteria have a net negative charge, they still have localized regions of positive charge from surface proteins attracting them to the substrate, and chemical substances, such as EPS, helping them stay adhered Berthelot [20].

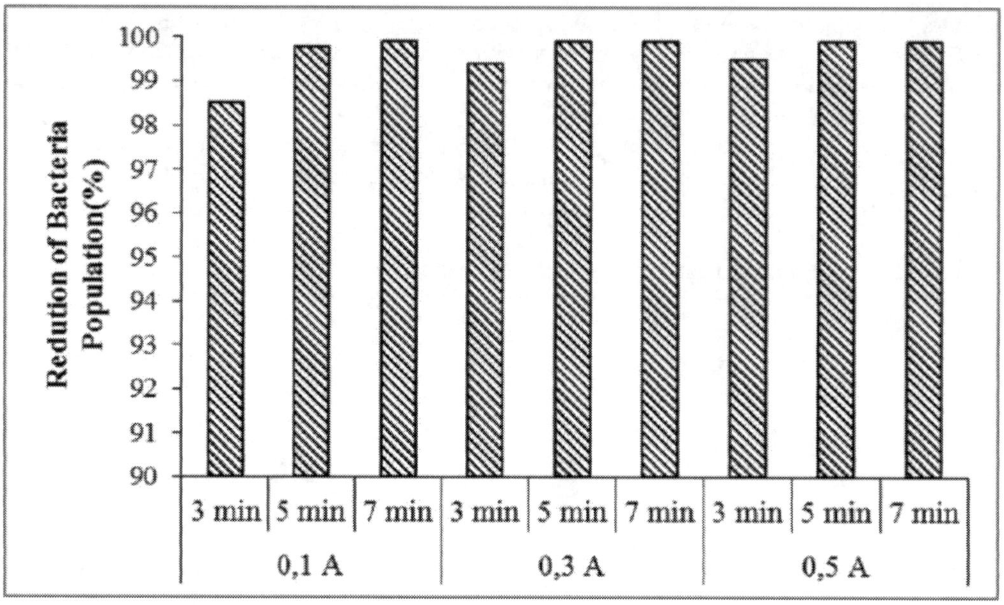

Fig. 3. Reduction Percentage of *Pseudomonas fluorescens* population.

4 Conclusion

The range of reduction of *Pseudomonas fluorescens* population was 98.5% to 99.9%, it indicated that the simple of ICAF system can reduce of *Pseudomonas fluorescens* population. The data will be used for the further investigation.

The author would like to thank KemenristekDIKTI through scheme of Penelitian Dasar Unggulan Perguruan Tinggi - PDUPT 2018 No. 890/PKS/ITS/2018 for funding this research.

References

1. M.C. Moura, E.V. Pontual, P.M.G. Paiva, L.C.B.B. Coelho. Méndez-Vilas A, Ed. Formatex. 11-22 (2013)
2. B. Marhaeni, O.K. Radjasa, M.M. Khoeri, A. Sabdono, D.G. Bengen, H. Sudoyo. J of Environ Protect **2(9)** 1245 (2011) doi: 10.4236/jep.2011.29143.
3. R.M. Donlan. Emerging Infectious Diseases **8(9)** 881-890 (2002)
4. M.E. Davey, G.A.O'toole. Microbiolo And Molecular Biol Rev **64(4)** 847–867 (2000)
5. G.J.C. Underwood, D.M. Paterson, R.J. Parkes. Limnol and Oceano **40** 1243-1253 (1995)
6. N. Muthukumar, A. Rajasekar, S. Ponmariappan, S. Mohanan, S. Maruthamutu, S. Muralidharan, P. Subramanian, N. Palaniswamy, N. Raghavan. Indian J of Experimen Botany **41** 1012-1022 (2003)
7. J.P.Black, T.E. Ford, R. Mitchell. Corrosion 88, National Association of Corrosion Engineers, Houston, Texas. (1988)
8. D.E. Nivens, P.D. Nichols, J.M. Henson, G.C. Geesey, D.C. White. Corrosion **42** 204-210 (1986)

9. C.O. Obuekwe, D.W.S.Westlake, J.A. Plambeck. Canadian J of Microbiol **33** 272-275 (1987)
10. A.J. Spiers, J. Bohannon, S.M. Gehrig, P.B. Rainey. Molecular Microbiol **50(1)** 15-27 (2003)
11. M.M. Baum, A. Kainović, T. O'Keeffe, R. Pandita, K. McDonald, S. Wu, P. Webster. BMC Microbiol **9(103)** 1-13 (2009)
12. W. Raza, W. Yang , Y. Jun, F. Shakoor, Q. Huang, Q. Shen. Carbohydr Polym **1;90(2)** 921-929 (2012) doi: 10.1016/j.carbpol.2012.06.021
13. H. Pratikno, H.S. Titah. Asian J of Appl Sci **9** 120-125 (2016)
14. H. Pratikno, H.S. Titah. Int J of ChemTech Res **9(12)** 600-609 (2016)
15. American Society for Testing and Materials. ASTM D1141-98 Standard Practice for Artificial Sea Water ASTM Society (2004)
16. H. Pratikno, H.S. Titah. *IPTEK,* The J for Technol and Sci **28(2)** 55-58 (2017)
17. J.P. Harley, L.M. Prescott. *Laboratory exercises in microbiology*. Fifth Edition McGraw–Hill Companies Texas. (2002)
18. H.S. Hong, J. Jeong, S. Shim, H. Kang, S. Kwon, K. Ahn, J. Yoon. Biotechnol Bioeng **100** 379-386 (2008)
19. M. Gabi, L. Hefermehl, D. Lukic, R. Zahn, J. Voros, D. Eberli. Urological Res **39** 81-88 (2011)
20. R. Berthelot. *Electrotaxis and Electrical Property Quantification of Pathogenic Bacteria Using Microfluidics and Nanoscale Imaging*. Thesis. Master of Science in Biophysics. University of Guelph (2017)

A comparison of the water quality between the mangrove areas Mulyorejo and Wonorejo

Luc Kauhl[1,3] , *Kayleigh* Lambregts[1,3,*] , and Suntoyo[2]

[1]Exchange student, Ocean Engineering Department, Faculty of Marine Technology, Institut Teknologi Sepuluh Nopember (ITS), Surabaya, Indonesia (60111)
[2]Associate Professor, Ocean Engineering Department, Faculty of Marine Technology, Institut Teknologi Sepuluh Nopember (ITS), Surabaya, Indonesia (60111)
[3]Student, Delta Acadamy, HZ University of Applied Sciences, Edisonweg 4, 4382 NW Vlissingen, The Netherlands

Abstract. The East Coast of Surabaya consists of large mangrove and fish pond areas. These areas are crucial for the ecosystem and for providing food and shelter for the local community. If the water quality decreases the mangroves will not be able to provide these functions. The water quality of these areas can be measured in two ways: by looking at the physico-chemical parameters or by looking at the biological parameters. Both offer a very good view of what state the area is in. The purpose of this study is to find out what the differences are between the Mulyorejo and the Wonorejo areas to see what differences there are within the East Coast of Surabaya and what could be the main cause of these differences. The data shows that the physico-chemical data does not comply with the limitation factors. The main problem are the nutrients which will have to be lowered to preserve the ecosystem. In terms of the biological state of the mangrove, there are few problems found. It is, however, unclear why the tree density in one of the Wonorejo areas is declining and it should be determined what the reason for this is.
Keywords: Mangrove, Fish ponds, Wonorejo, Mulyorejo, Physico-Chemical, Biological

1 Introduction

Mangroves are an important part of the coast line of Surabaya. They provide an ecosystem that protects the people and offer a source of food for both humans and aquatic species. The East Coast of Surabaya contains a large mangrove area. This mangrove area is divided in multiple parts which are all influenced by different rivers. The area is part of a development plan in which this area is meant as a place where the main functions are to protect and rehabilitate the marine environment, develop nature tourism as well as fishing and fishery cultivation areas.

The research is focused on two areas. The first area is the Mulyorejo area, which has a small river flowing through it and which is mainly influenced by domestic waste from

[*] Corresponding author: kayleighlambregts@hotmail.com

the city of Surabaya. The second area is the Wonorejo area, this area has a large river flowing through it, the Brantas river. This river flows through most of East-Java and brings in a lot of pollutants from its catchment area which makes it unclear where the main pollution is coming from even though there is extensive research being done in this area. Both areas are also greatly influenced by illegal logging for the creation of fish ponds and residential areas.

In this paper we aim to identify, analyze and compare these mangrove areas and fish ponds to determine what influences the urban runoff and domestic waste from the city of Surabaya bring into the system, and to see what influence the fish ponds and mangroves have on the water quality in these areas.

2 Theoretical background

2.1 Mangroves and fish ponds

Mangrove forests have adapted to the low oxygen and high salinity of the coastal zones. They limit their salt intake and their water loss. Next to that, they also stabilize the coastline, reduce erosion from storms, currents, waves and tides. Their root system allows them to live in this area. These roots also make an attractive living environment for many aquatic species as a breeding and feeding, and shelter area [1].

This also makes the area surrounding the mangroves interesting for fish ponds, these fish ponds need to receive an effluent or sludge. The fish can then feed on algae and other organisms but can also filter nutrient rich water from the mangroves. Fish that live in aerobic ponds can effectively reduce algae and help control mosquito populations. The fish do not improve the water quality a lot but due to their economic value can offset the cost of operating a treatment facility [2].

2.2 Coastal zone

The Indonesian coastal zone has a high potential to become the most important coastline in the world in terms of economic activity. The coastal zone also holds one of the richest ecosystems in the world, characterized by extensive mangroves, coral reefs and seagrass beds. These ecosystems play an important role in determining system response to environmental changes [3]. The mangroves represent the main ecosystem in marine and brackish coastal zones and contain large amounts of resources which support the major fisheries and mariculture activities [4].

2.3 Water quality

Water quality is defined in terms of the chemical, physical and biological contents of water [5]. The definition of water quality assessment has reached a level of complexity which requires simultaneous consideration of multiple aspects. This calls for a multi-dimensional approach to an inevitable necessity to assess the water quality. To measure the water quality, theoretically speaking, there are an infinite amount of ways in each kind of water body, every different pollutant, water use and combinations of these depending on the objective. In practice, however, the amount of different measuring methods is limited due to influences from the environment, equipment, et cetera and are defined in different types (e.g., trend monitoring, basic survey, etc.) [6].

2.3.1 Physico-chemical

The physico-chemical parameters have a great influence on the aquatic system. Some of the most influential parameters being temperature, rainfall, pH, salinity, and dissolved oxygen. These factors have a limiting factor or the survival of aquatic organisms, such as flora and fauna. They can be altered by a variety of factors such as municipal effluents and industrial discharges [5].

2.3.2 Biological

Exposure of a biological community acts as an integrator of present and past environmental effects, this attribute makes them useful as indicators for the measurement of system changes. The response of these communities is well predictable and based on the indicator concept. This concept is well founded as observations that organisms reflect their environment are found in writing dating from as early as Aristotle [7].

3 Materials and method

This methodology is based on the methodology used by the Environmental Agency of Surabaya and describes how the data from the Mulyorejo area has been collected and analyzed to be compared with data that has been provided by the Environmental Agency.

The physico-chemical parameters pH (pH-meter), temperature (thermometer), turbidity (nephelometer), total suspended solids (gravimetric), salinity (salinometer), dissolved oxygen (iodimetric), biological oxygen demand (Winkler method), chemical oxygen demand (reflux), total ammonia (spectrophotometric), total phenol (spectrophotometric), phosphate (spectrophotometric) and nitrate (spectrophotometric) have been measured at 4 (four) sampling points in the Mulyorejo area and have been analyzed by the Environmental Engineering lab. The methodology of the lab analysis is described in the 22nd edition of the book Standard Methods for Examination of Water and Wastewater (APHA) by Rice et al. [8].

The biological method consists of the analysis and determination of the vegetation. The vegetation has been assessed by the determination of the Important Value Index (trees: given in Eq. (1); seedlings, saplings and other vegetation: given in Eq. (2), the Shannon Wiener index given in Eq. (3) and the Evenness index given in Eq. (4), respectively.

$$INP = KR + FR + DR \tag{1}$$

$$INP = KR + FR \tag{2}$$

$$H' = \sum[p_i \ln (p_i)] \text{ with } p_i = (\frac{n_i}{n}) \tag{3}$$

$$E = \frac{H'}{\ln(S)} \tag{4}$$

This analysis has been done along two transect lines consisting of two plots each in the Mulyorejo area. The analysis was done using the following equipment: GPS, measuring tape, rope, preprinted data sheets and pencils.

The two mangrove location points in the Mulyorejo area are found in the mouth of the river and in the middle part of the river. The two fish pond samples were taken on opposite sides of the river close to the river bank. The Biological sample was taken in the bend of the river opposite each other. The Wonorejo area is located in the south part of the coast (Figure 1).

Fig. 1. The sampling locations

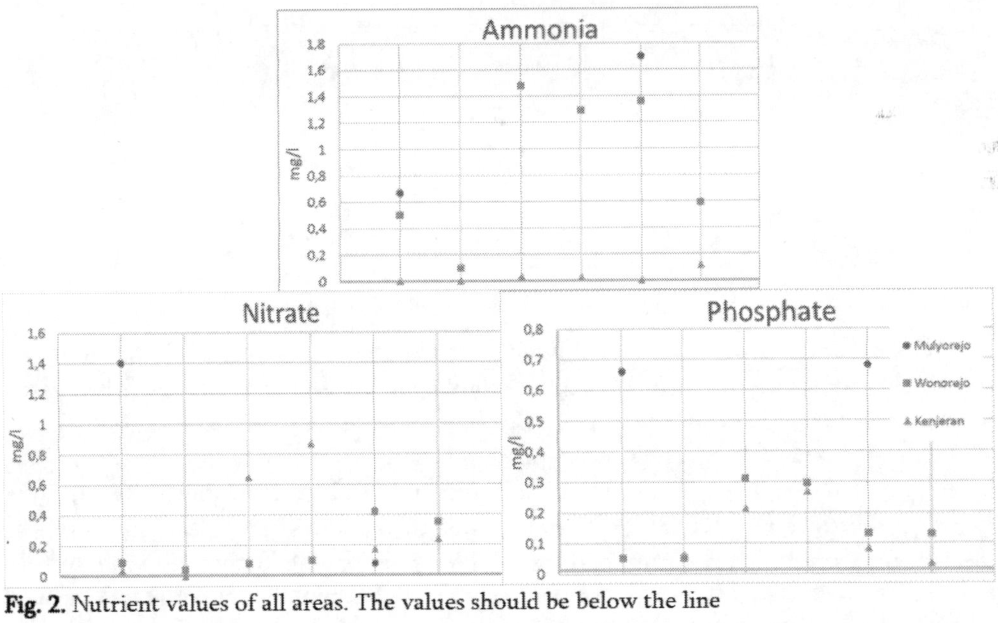

Fig. 2. Nutrient values of all areas. The values should be below the line

After each sampling trip, the data was formatted using excel. The data was labelled with a full description of the sample collection procedure including the location, type of sampler, quantity sampled, number of samples, and types of filtration, apparatus and filters. It was then formatted to have the same format as the provided data for analysis.

4 Results and Discussion

The nutrients are divided in three different parameters; Nitrate, Ammonia and Phosphate (Figure 2). In the Mulyorejo area the nutrient levels are the highest of the three areas and are fluctuating quite severely between the two sample points. The conditions in Wonorejo and Kenjeran are similar but are still higher than the limitation factor.

When looking at the physical parameters, it becomes apparent that turbidity and total suspended solids exceed the limitation factors (Figure 3). Temperature has no limitation factor but has a value that is to be expected in the local climate. The temperature of the samples taken from Mulyorejo is most likely lower because the location and date differ from the other areas.

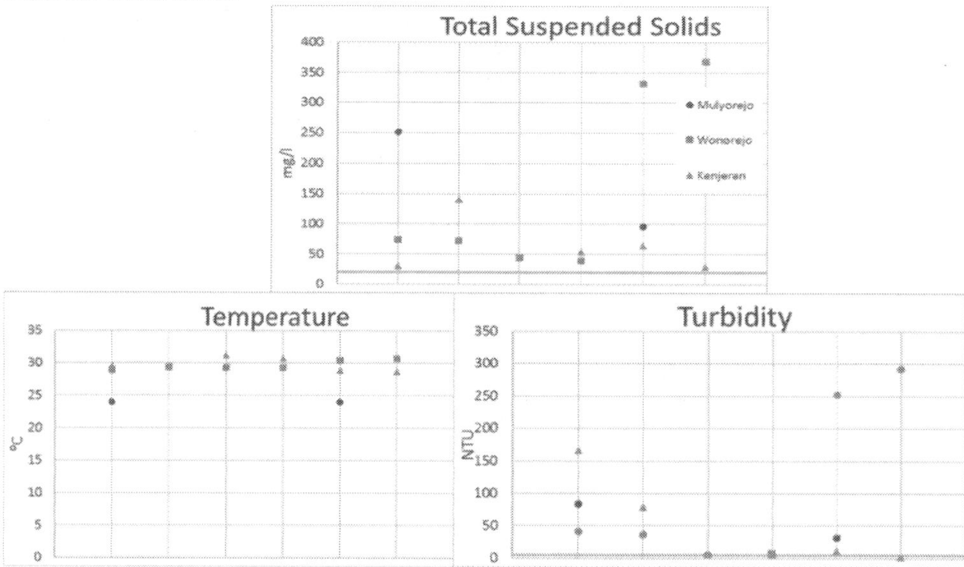

Fig. 3. Physical parameters. The values should be below the line

The BOD and DO conditions of Wonorejo are in accordance to the limitation factor and have a stable trend (Figure 4). The first sample point of DO in Wonorejo is below the limitation factor which is most likely caused by a divergence of the standard deviation. Both Mulyorejo and Kenjeran exceed the limitation factors of BOD and DO. When looking at BOD there are many fluctuations visible within the locations. Research by Sakinah *et al* [9] shows that in the Wonorejo area these values have great impact on the local system, with values corresponding to the values found in this research. While, Suntoyo and Sakinah [10] have shown that the phenol concentration in northern Wonorejo estuary is higher than other areas based on the numerical modelling simulation.

Salinity does not have any limitation factors but has very significant fluctuations in the different sample points within areas, ranging from almost 0 to 45 ppt (Figure 5). The

Ph should be between 7-8.5 to be in accordance to the limitation factor. As Figure 5 shows, almost all samples lie within this range.

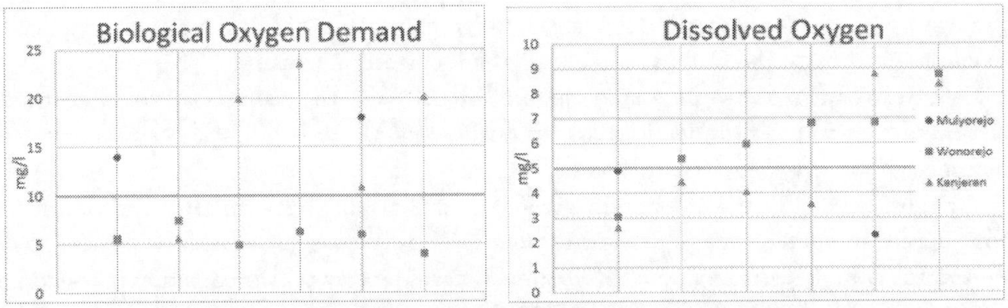

Fig. 4. BOD and DO. The value of BOD and the value of DO should be below the line

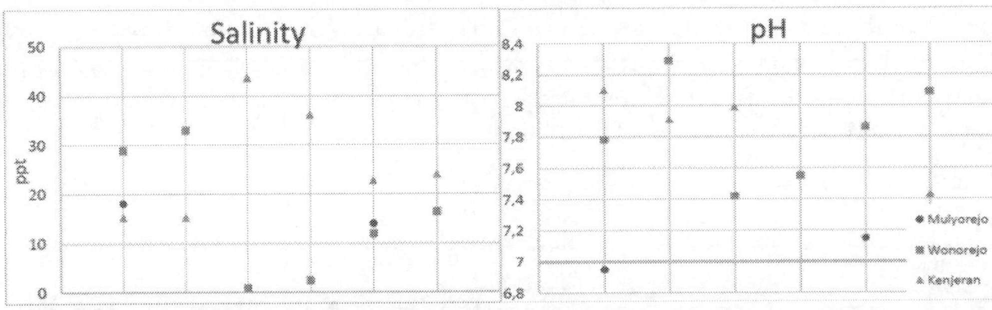

Fig. 5. Salinity and pH. The value of pH should be between 7 and 8.5

Table 1 describes the physico-chemical parameters of the fish ponds in the Mulyorejo area. When comparing it to the mangroves in the Mulyorejo area, it becomes apparent that there is little difference between the parameters measured. The nutrients are over the limitation factor, and the same goes for the physical, biological oxygen demand and the dissolved oxygen condition.

Table 1. The physico-chemical values of the fish ponds in the Mulyorejo area

Test Description	Unit	Regulatory Limit*	fish pond 1	fish pond 2
Temperature	^0C	Natural	25	24
Turbidity	NTU	5	**64**	**18,4**
TSS (Total Suspended Solids)	mg/l	20	**96**	**72**
pH	-	7-8.5	7,3	7,05
Salinity	ppt	Natural	20,7	20,9
DO (Dissolved Oxygen)	mg/l	>5	**4,6**	**3,6**
BOD_5 (Biological Oxygen Demand)	mg/l	10	**16**	**12**
Free Ammonia (NH_3-N)	mg/l	Nihil	**0,77**	**0,73**
Phosphate(PO_4-P)	mg/l	0,015	**0,86**	**0,44**
Nitrate (NO_3-N)	mg/l	0,008	**0,1**	**0,07**

In the Kenjeran area, the most apparent thing is that area 1 in terms of density (Ka) is considered as good to very dense (density is above 1500 trees/ ha; Regulation of Environmental Minister No. 201/2004) while the density of area 2 is considered as damaged to rare (density is lower than 1000 trees/ ha) (Table 2). The most important species according to the INP in both areas is the Avicennia marina.

When looking at the saplings and seedlings in this area, the numbers of these are rising in comparison to the number of trees. In both areas, there is one species found in this category.

When looking at the Wonorejo areas 1 and 2 it can be seen that the density is considered as good to very dense (Table 3). The Wonorejo area 2 however, has a substantially higher density. Wonorejo area 3 has a density that is considered as good to moderate (density is between 1000-1500). In the Wonorejo areas, there are multiple species present but when looking at the INP it can be said that the Avicennia marina is the most important and abundant species in these areas as well.

When looking at the saplings and seedlings in this area, the amount of species found is decreasing. The absolute number of seedlings and saplings is however increasing in comparison with the number of trees in the Wonorejo areas 1 and 2. In Wonorejo area 3 there is a decline in the number of seedlings.

Table 2. The governmental data from Kenjeran; this is the coastal part of the Mulyorejo area with Ka in trees/ ha

Kenjeran 1 Species	ni	Ka	INP	Kenjeran 2 Species	ni	Ka	INP
Trees:				*Trees:*			
Avicennia marina	55	1833	259,16	Avicennia marina	25	833	300
Avicennia alba	6	200	40,83				
Total	61	2033	300	Total	25	833	300
Saplings:				*Saplings:*			
Avicennia marina	30	4000	200	Avicennia marina	19	2533	200
Total	30	4000	200	Total	19	2533	200
Seedlings:				*Seedlings:*			
Avicennia marina	11	9167	200	Avicennia marina	40	33333	200
Total	11	9167	200	Total	40	33333	200

In the Mulyorejo areas the density is considered as good to very dense with only one species present. The number of saplings and seedlings in these areas are also substantially higher than the number of trees found (Table 4).

Table 3. The governmental data from the Wonorejo area; Ka is in trees/ ha

Wonorejo 1 Species	ni	Ka	INP	Wonorejo 3 Species	ni	Ka	INP
Trees:				*Trees:*			
Avicennia marina	46	1533	230	Sonneratia alba	3	100	36
Rhizophora stylosa	2	67	21	Avicennia marina	17	567	115
Xylocarpus moluccensis	3	100	26	Rhizophora stylosa	8	267	63
Excoecaria agallocha	2	67	23				
Total	53	1767	300	Total	40	1333	300
Saplings:				*Saplings:*			
Avicennia marina	12	1600	48	Sonneratia alba	2	267	35
Rhizophora stylosa	46	6133	117	Avicennia marina	11	1467	77
Xylocarpus moluccensis	4	533	35	Avicennia alba	4	533	44
Total	62	8267	200	Rhizophora stylosa	4	533	44
Seedlings:				Total	21	2800	200
Avicennia marina	4	3333	42	*Seedlings:*			
Rhizophora stylosa	19	15833	158	Rhizophora stylosa	1	833	200
Total	23	19167	200	Total	1	833	200

Wonorejo 2 Species	ni	Ka	INP
Trees:			
Avicennia marina	94	3133	300
Total	94	3133	300
Saplings:			
Avicennia marina	45	6000	200
Total	45	6000	200
Seedlings:			
Avicennia marina	3	2500	110
Rhizophora stylosa	2	1667	90
Total	5	4167	200

Table 4. The data from the researched area; this area is more inland then the other two areas. Ka is in trees/ ha

Mulyorejo 1 Species	ni	Ka	INP	Mulyorejo 2 Species	ni	Ka	INP
Trees:				*Trees:*			
Avicennia alba	62	3100	300	Avicennia alba	52	2600	300
Total	62	3100	300	Total	52	2600	300
Saplings:				*Saplings:*			
Avicennia alba	29	5800	200	Avicennia alba	37	7400	200
Total	29	5800	200	Total	37	7400	200
Seedlings:				*Seedlings:*			
Avicennia alba	23	28750	200	Avicennia alba	25	31250	200
Total	23	28750	200	Total	25	31250	200

5 Conclusion

When looking at the nutrient concentrations, it is clear that the amount of nutrients is too high in this ecosystem. This stimulates the algae blooms, which influences all aquatic life such as fish and invertebrates, lowering the biomass of the system in the long term. The most probable cause of these high nutrient amounts is domestic waste and urban runoff.

Turbidity and Total Suspended Solids have large fluctuations in their concentrations. They often do not comply with the limitation standards. This could be caused by the sedimentation of the water or the erosion and tidal waves. This causes the water to become more turbid and increases the amount of particles in the system. This makes finding the point source difficult as there are many influences. The Temperature does, in this case, not tell much about the water quality. It is in this area mostly influenced by the local climate and not by pollution.

The BOD and DO are linked to each other, when BOD becomes higher, DO will decrease. BOD is in term greatly influenced by the amount of nutrients present in the system as these nutrients often cause the number of algae to increase. When BOD is high, then this influences the populations of fish and all other aquatic organisms dependent on oxygen. The low concentration of DO might cause the death of some aquatic life.

When talking about the pH it can be said it is neutral, from this we can say that the CO2 value is also around an average value. The values of Salinity are most likely fluctuating due to the large influences from the sea at high tide and from the river at low tide. The values of Salinity are as expected in this area.

The conditions in the fish ponds do not differ much from the mangroves. This is not in line with the expectations as fish ponds are expected to have a cleansing function.

The conditions of the biological parameters are overall according to the regulations. The Wonorejo area has the highest species diversity while the Mulyorejo area has the highest tree density of all the areas. When looking at the seedlings and saplings it can be concluded that the density in all areas will become higher in the future except for the Wonorejo area 3. The reason that there are only one or two species found in some of the areas is due to the fact that the genus Avicenniaceae is a pioneer in mangrove ecosystems. This genus traps the sediment which will allow other species to grow in the future.

Following these results, the next step is to extend the research to find out which nutrient has the biggest influence on this aquatic system to be able to find a solution to limit the influence of these excess nutrients. There is also need to make sure that the trend of the increasing density of the mangrove continues, there should be regular monitoring both in the coastal area and in the inland part of the mangrove. It is also important to find out why in the Wonorejo 3 area the number of seedlings and saplings found is so low to avoid this happening in other areas. And lastly, to prevent problems in the future, the amount of domestic waste coming into the system should be lowered. It is therefore important to find the source of this waste and to find a solution to limit it.

We would like to thank our supervisor Dr. Suntoyo (Associate Professor, Ocean Engineering Department, Faculty of Marine Technology, Institut Teknologi Sepuluh Nopember) who has enabled us to do this project and to present it at the ISOCEEN congress. Secondly, we would like to thank Muhammed Romadhoni, S.Si for helping us during this research, as the contact person of the Environmental Agency of Surabaya and for providing us the biological and chemical data from the mangrove areas. At last we would like to extend our gratitude to the research group Building with Nature (HZ University of Applied Sciences) for providing us with a budget for the chemical and biological analysis of the samples.

References

1. NOAA. What is a "mangrove" forest? Oceanservice.noaa.govAvailable from: https://oceanservice.noaa.gov/facts/mangroves.html. 2017
2. Eawag. Fish Pond (Aquaculture) | The online Compendium of Sanitation Systems and Technologies [Internet]. Ecompendium.sswm.info. 2017 [cited 10 September 2017]. Available from: http://ecompendium.sswm.info/sanitation-technologies/fish-pond-aquaculture
3. S. Sukardjo. Oceanographic Literature Review. **42**(9):776. (1995)
4. S. Sukardjo. Kyoto University. (2002).
5. E. Lawson. Advances in Biological Research. **5**(1):8-21. (2011)
6. L. Chapman, C. Chapman, M. Chandler. Biological Conservation. **78**(3):263-270. (1996)
7. P. Asokan. Mangroves and its importance to Fisheries - CMFRI Repository [Internet]. Eprints.cmfri.org.in. 2012 [cited 3 October 2017]. Available from: http://eprints.cmfri.org.in/9774/
8. E.W. Rice, R. B. Baird, A. D. Eaton, L. S. Clesceri, Standard Methods for the Examination of Water and Wastewater, 22nd edition. Washington, DC: American Public Health Association. (2012)
9. W. Sakinah, Suntoyo, Mukhtasor. Appl. Mechanics and Materials. **862**:96-101. (2017)
10. Suntoyo, W. Sakinah. Modelling of Phenol Contamination in Wonorejo Estuary, Indonesia. *IOP Conf. Ser.: Earth Environmental Science.* **79** 012005. (2017)

Domestic Waste Mass Estimation and Perceptions of Coastal Citizen about Waste Management in Besuki Region, Situbondo, East Java, Indonesia

Wazirotus Sakinah[1,*], *Dina T* Septiningtyas[1] , and *Anita D* Pahlewi[1]

[1]Universitas Abdurachman Saleh, Situbondo, 68312, Indonesia

Abstract. Coastal citizen in Besuki region have a habit of throwing garbage directly into estuary and sea. To know the bad conditions that can occur in the marine environment especially in Besuki region so domestic waste mass estimation is needed. Proportional random sampling is used to determine the number of samples and measurement of domestic waste is conducted for 8 days. Measurement with SNI (Indonesia National Standard) Number 19-3964-1994 method is used. To know perceptions of coastal citizen about waste management, the distribution of questionnaires to the sample was done. The result is waste mass estimation in Besuki Coastal Region is 167.57 g/people.day where non-organic waste more than organic waste that has percentage 63% and 37%, respectively. From questionnaire survey, 73% citizen throw their wastes in the sea or/and the river, 27% has been burned, and 0% in landfill. Generally, 80% citizens do not have the basic knowledge about marine debris impact and 20% citizens have knowledge about marine debris impact. Their interest about self-waste management is 33%.

1 Introduction

Indonesia as the second largest contributor of plastic waste in the world is the main topic of the conversation of the world community today. Every year Indonesia donates about 3.2 million tons of plastic waste [1]. On March 2018, plastic waste in Jakarta bay became a big issue in Indonesian news. The thickness of waste in Jakarta bay about 1 meter above the sea surface level. The Indonesian Biodiversity Foundation (Kehati) predicts that the plastic waste that is constantly polluting the Jakarta bay area will be a ticking time bomb if not handled quickly [2]. Adding the facts that waste pollution become an urgent problem in Indonesia. Waste pollution cannot separate from plastic waste. Plastic waste is generally discarded in waters from the river that will eventually empty into the ocean and directly discharged into the sea. Similarly, the coastal citizens in Besuki region generally have a habit of throwing garbage directly into estuary and

* Corresponding author: wazirotus.sakinah@gmail.com

sea. Once introduced into rivers, plastic debris may sink, be deposited on river banks and/or undergo transfer to the marine environment [3]. Plastic debris cannot decomposed easily. In a long time, it transform to microplastic, it is not dissapear from the environment. It can bring negative impacts on many aspects, not only endanger the livelihood of people but it also threatens to poison marine ecosystems.

Marine ecosystems in Indonesia have been much damaged. Mangrove area in Indonesia until 2010 from Ministry of Environment Indonesia has percentage 56.91% in good condition, 10.69% medium, 7.20% damaged, and 25.20% unidentified. As for coral reef data from the results of oceanographic research LIPI in 2012 showed only 5.30% in very good conditions, 27.19% good, 37.25% good enough, and 30.45% less good [4]. A previous study between 2011 and 2014 found reefs near Indonesia were littered with the most plastic, while the lowest concentration was found in Australia [5]. Similarly, the seagrass condition also has been much damaged. Seagrass beds are vulnerable to natural disturbance and human activities, including the agricultural waste disposal, oil spill, and organic waste [6]. The impact of waste pollution also bring the effect on animal welfare. As example crocodile with tyre choker found in Palu, Central Sulawesi. Another case happened in Makassar, anthropogenic debris was found in 28% of individual fish and in 55% of all species. All of the anthropogenic debris recovered from fish in Indonesia was plastic [7].

Previous research examined composition of municipal solid waste (MSW) during the period of 2000 – 2006 in Taiwan. There are increasing amount of MSW in Taiwan from year to year [8]. Due to this phenomenon proper waste management is needed as a public benefit and obligation. Improper waste disposal by one individual affects the entire citizen [9]. But, as in many regions in Indonesia, waste management is a complex issue. Generally, citizen have a poor waste management, throw their wastes in the waters is the most treatment that always happen in Besuki coastal region. The highlights of this current study to investigate how much waste that produces per day in this region and perceptions of citizen about waste management in their place.

The paper is organized as follows. It begins by preliminary survey for marine debris in Besuki coast, listing the methods used to examine debris mass and types, and citizen perceptions about debris disposal practices and the factors.

2 Study Area

The study area is located in Pesisir village, Besuki region, Situbondo, East Java, Indonesia. Geographically located in 7.725 South latitude and 113.686 East longitude.

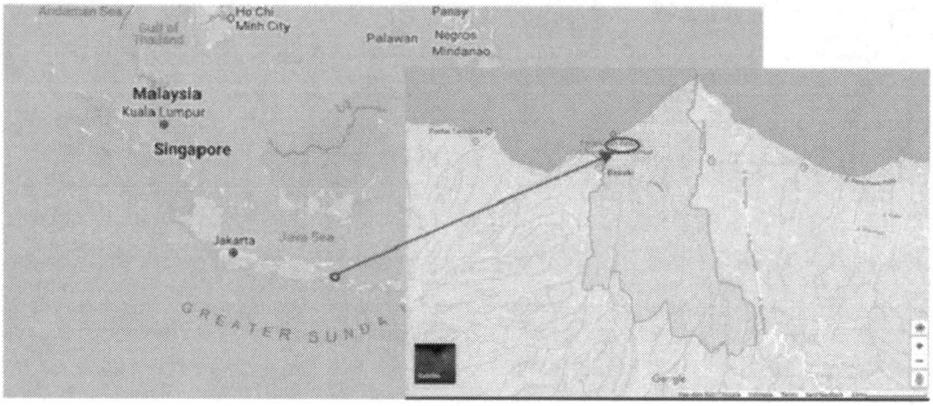

Fig. 1. Study area

3 Material and Method

This research did two kinds of method, field measurement with proportional sampling method and giving questioner to some citizen as sample. Field measurement was done at October 2017. Sample determination based on proporsional random sampling, ie the samples were taken randomly on each unit of the settlement area with the number of different samples in each unit of the settlement. Sampling was did in 8 (eight) days respectively at the same location with sampling time 24 hours for each sample [10]. Digital scale is used to measure domestic waste mass. The formula to measure the estimation of domestic waste mass is:

$$= \frac{\left(\frac{B_{S1}}{u}+\frac{B_{S2}}{u}+\cdots+\frac{B_{Sn}}{u}\right)}{Sn} \quad \text{g/people/day} \qquad (1)$$

where:
B_s = measure of waste mass (g/day)
u = unit of waste producer (people)
Sn = number of sample (people) [3]

Questionnaires are used to obtain perceptions of coastal citizens about waste management. The questionnaire contains several questions covering how citizen do waste management, are they do the sorting of waste, their knowledge of marine debris impact, and their interest in self-management waste.

4 Result and Discussion

4.1 Domestic Waste Mass in Besuki Coastal Region

Generally, up to 80% marine debris comes from land-based sources, which is transported via sewage/drainage system, natural waterways, wind or human neglect [11, 12]. Mostly, the disposal is plastic including plastic garbage bags and other plastic products. This statement can be proven from our preliminary survey in Figure 2.

Fig. 2. Preliminary condition in Besuki region, (a) garbage in river, (b) garbage in coastal area

Plastic, the most marine debris in Besuki region, was known from Figure 2, is the difficult waste to decompose. For example plastic bottles, bottled water is carried on board for drinking and/or cooking purposes and plastic bags that are commonly used to pack goods or other types of waste [12].

Based on data from the statistical center agency, population in Besuki is 64,655 people with coastal region has 10,021 citizen, where the average household size was 5 people. The results based on sampling data collection methods can be seen in Table 1.

Table 1. Domestic waste mass from sampling data collection in Besuki coastal region.

Sampling day-	**Total mass (g/household.day)**		**Total (g) waste/day**	**Total (g) waste/unit/day**
	Organic	non organic		
1	221,7	1224,0	1445,7	289,14
2	469,9	649,0	1118,9	223,78
3	247,8	882,5	1130,3	226,06
4	589	744,9	1333,9	266,78
5	565,1	660,7	1225,8	245,16
6	219,4	295,5	514,9	102,98
7	507,2	358,3	865,5	173,1
8	298,7	444,8	743,52	148,704
Average				1047,315
Total				1675,704
Estimate g/people.day				167,5704

Throughout sampling data collection, the domestic waste mass in Besuki region are always changing. The waste mass in each household is 1,047.32 g/day so the estimation waste mass in Besuki coastal region is 167.57 g/people.day. If this amount is multiplied by the number of inhabitans in Besuki coastal region, then the amount will be obtained 1679 kg/day or 1,68 tons of garbage every day in that region. The fact is, most of the waste end as marine debris as on Figure 2. Such as previous research, most of municipal solid waste still dumped or landfilled, especially in countries with low income per capita and globally just about 15% is recycled [13]. Generally, domestic waste from Besuki coastal region has non-organic waste more than organic waste, it can be seen on Figure 3.

Organic waste and non-organic waste from sampling data collection has percentage 73% non-organic waste including papers and plastics which have percentage 33% and 40%, respectively and 27% organic waste such as food garbage from total waste sample in 8 days. Decompose of organic waste such as fermentation for food waste need 4 days with act in anaerobic bacteria [14]. But, decompose non-organic waste need longer time. Plastic bags, the common item in Besuki coastal region domestic waste, take 10-1000 years to decompose. Another common item is paper waste, it takes 2-6 weeks to get completely decomposed and fruit peel takes months to decompose [15]. The long time it takes waste to decompose causes the build-up of waste, especially in the river and sea in Besuki coastal region.

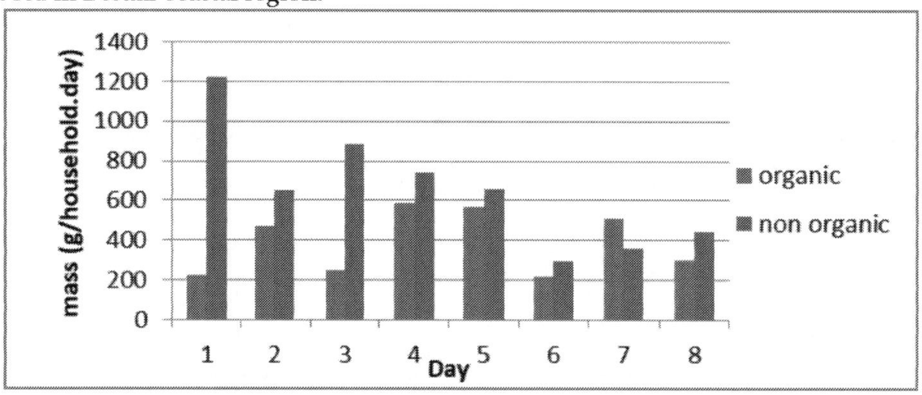

Fig. 3. Comparison of organic and non-organic waste in Besuki coastal region

4.2 Perceptions of Besuki Coastal Citizen about Waste Management

A total 4 questions for each sampled households respondents was used for the survey questionnaire. The questionnaire is multiple choice questions because some respondents are illiterate. The first question about waste management. 73% respondents throw their waste in the sea or river and 27% waste are burned, see Figure 4.

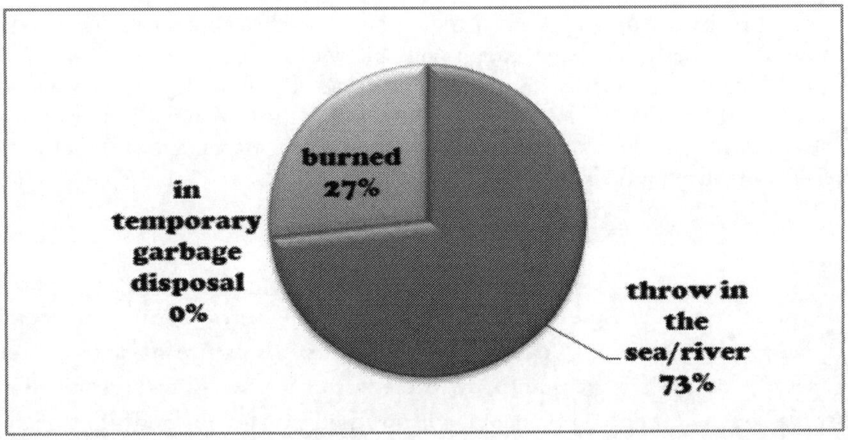

Fig. 4. Waste management diagram in Besuki coastal region.

Generally, the reason why they usually throw their waste to the waters is waste can flow straight away. The reason why wastes are burned is another people throw the waste near the sample house. There are no one throw their waste in landfill because there is no landfill in this region. This waste management can be considered as a traditional common. A traditional commons that seeing the ocean as a dumping ground for waste make a tragedy in the global oceans. Absenting a governance system to impose rules for behaviour and the rational individual sees that the shared cost of polluting are often smaller than personally paying for the processing of their waste [16]

The second question about sorting waste. 87% respondents said they do not sort their waste because there is no place and they do not know why sorting waste is needed. 13% respondent sort their waste because some waste can be sold. Generally, they do not know about marine debris impact. From all respondent, only 20% know about the impact of marine debris. It is make them to have a bad habit to throw their waste to the waters. Finally, they do not interest about self waste management. It can be seen from the questionnaire result, only 33% said they interest about self waste management. Lower middle countries has percentage 49% waste is dumped, 11% landfilled, 2% composted, and just 5% is recycled [13].

Table 2. The result of questionnaire

No	Descriptions	Percentage
1.	Waste Management	
	- Throw in the sea/river	73%
	- Burned	23%
	- Landfill	0%
2.	Sorting waste	
	- Yes	13%
	- No	87%
3.	Knowledge on marine debris impact	
	- Yes	20%
	- No	80%
4.	Self waste management interest	
	- Interest	33%
	- No interest	67%

The third question about their knowledge on marine debris impact. 80% respondents does not know about marine debris impact. It makes them pay less attention to their environment. It is the reason why Besuki region has a poor waste management today. The last question about the respondents interest on self waste management. 67% respondents does not have interest on self waste management. Generally, they does not know why they need to manage their waste. This question have correlation to the third question. Poor knowledge about marine debris impact make a poor waste management. Finally, they are not interested in managing their waste independently.

5 Summary

The waste in Besuki coastal region was dominated by non-organic type which mass in each household is 1,047.32 g/day, while the estimation waste mass in its region is 167.57 g/people.day. 73% of people throw their wastes in the sea or/and the river, 27% of wastes was burned, and 0% of people throw their wastes in landfill, it means that they have a bad habit of disposal debris. Generally 80% of them do not have the basic knowledge about marine debris impact. While, based on the survey, 33% of them has a good interest about self-waste management. So, from the data above we can conclude that Besuki Coastal Region has a poor waste management.

References

1. Information on http:// www.cnnindonesia.com
2. D. Prasongko, *Plastic pollution; ticking time bomb in Jakarta Bay* (Tempo.co, 2018)
3. J. Gasperi, R. Dris, T. Bonin, V. Rocher, B. Tassin, Env. Pollut. **195**, 163 (2014)
4. B. Pramudyanto, Jur. Ling. Wid. **1:4**, 21 (2014)
5. M. Ompusunggu, *Study to challenge claim Indonesia second-biggest marine polluter* (The Jakarta Post, 2018)
6. Greenpeace Southeast Asia, *Greenpeace laut Indonesia dalam krisis* (Greenpeace Southeast Asia, Jakarta)
7. C. M. Rochman, A. Tahir, S. L. Wiliams, D. V. Baxa, R. Lam, J. T. Miller, FC. Teh, S. Werorilangi, S. J. Teh, *Anthropogenic debris in seafood: Plastic debris and fibers from textiles in fish and bivalves sold for human consumption* (Scientific Reports, 2015)
8. W. T. Tsai, Bior. Tech. **99**, 5309 (2008)
9. R. M. Yoada, D. Chirawurah, P. B. Adongo, BMC. Pub. H. **14:697**, 1 (2014)
10. SNI 19-3964-1994, *Metode Pengambilan dan Pengukuran Contoh Timbulan dan Komposisi Sampah Perkotaan BSN*
11. D. A. Cooper, P. L. Corcoran, Mar. Pol. Bul. **60**, 650 (2010)
12. C. L. Chen, T. K. Liu, Mar. Pol. **40**, 34 (2013)
13. H. Jouhara, D. Czajczyńska, H. Ghazal, Krzyżyńska, L.Anguilano, Energy **139**, 485 (2017)
14. R. Slezak, J. Grzelak, L. Krzystek, S. Ledakowicz, Was. Man., (Article in Press) (2017)
15. W. Rathje, C. Murphy, *Rubbish! The Archaeology of Garbage* (The University of Arizona Press, Tucson, 2001)
16. L. L. Micah, Mar. Pol. Bul. **127**, 310 (2018)

Tidal, wave, current and sediment flow patterns in wet season in the estuary of Porong River Sidoarjo, Indonesia

Engki A. Kisnarti[1,*], and *Viv* Dj. Prasita[1]

[1] Department of Oceanography, Hang Tuah University, Jl. Arif Rahman Hakim 150, Surabaya-Indonesia

Abstract. The objectives of this research are to analyze characteristics of physical oceanography, such as : tides, waves, currents, and discharges at Muara Kali Porong. This research also discuss sediment flow patterns and morphology in around the Estuary of Porong River. Tidal data were used as correction to the depth. The calculation to determine the tidal current velocity and wind data along with current data are used for simulation model. Sedimentation process with a simulation of 15 days in the West Season occured in the Northeast of Lusi Island with sediment thickness ranged from 1.6 to 2.6 m.

1 Introduction

The bursts of mud and hot mudflow in Porong, Sidoarjo that have occurred since May 29, 2006 until today still continues, and there are no signs that this natural phenomenon will stop in the near future. The amount of mud volumes emerging from the center of the eruption is enormous, in 2006-2007 estimated at 100,000 m^3/day and even reaching 180,000 m^3/day in December 2006, and tend to decrease to about 75,000 m^3/day in July 2009. The estimated volumes of bursts 50.000 m^3/day in September 2011. In 2006, the volumes of mud that went out continuously caused panic. With the rapid assessment of the government, it is determined that the mud should be discharged into the sea through the Porong River. This is in accordance with Presidential Regulation 14/2007 and its changes, the mud flow of Sidoarjo must be channeled into the sea through Porong River. [1].

The disposal of mud to Kali Porong has created disadvantages in various fields, both social, economic, and ecological. The mud flowing into Kali Porong adds to the adverse effects of human activities in the river catchment area on the ecology and biogeochemistry of the estuary and coastal waters of the Madura Strait [2]. The result of mud disposal into the estuary of Porong River are not only water quality changing. Remote sensing data is able to provide spatial and temporal information about other impacts of mud disposal to the Porong River. The results indicate that coastal mangrove areas have increased significantly due to Sidoarjo Mud (LUSI) - causing mudflow deposits near the mouth of the river [3].

* Corresponding author: engkiandri@hangtuah.ac.id

Hydrodynamic model can also be used to know the sediment flow pattern in Muara Kali Porong [4], [5], [6]. The general distribution of sediment distribution in Porong waters shows mud sediments found on the banks of rivers and sand sediments found in the middle of the river. The basic sediments of the waters around the Porong estuary are sand, clay and silt lemps clay [7].

Research on the characteristics gradually has also been done, namely about the morphology in Estuary of Porong River which tends to be influenced by the accretion and abrasion process, as well as the tidal and currents relations around the Estuary of Porong River [8], [9]. To complement the research that has been done, it needs a more in-depth study with the purpose to analyze the physical characteristics (tides, waves, currents, discharges) flowing into Muara Kali Porong, sediment flow patterns to morphology, and coastline changes in around the Estuary of Porong River for 10-15 years with satellite imagery in the western season.

2 Methodology

2.1 Study Area

Porong river is a tributary of Brantas river which originated in the city of Mojokerto (Lengkong Baru Dam), flows eastward and empties into the Madura Strait. The name Porong is taken from the name of a sub-district located at the southern end of Sidoarjo regency. Porong River has two tributaries, Sadar River with 406.70 km^2 of river flow that empties into Krembung Village and Kambing River with 196.60 km^2 of river flow. Geographically, Porong river lies between 112.5^0 BT- 112.9^0 BT and 7.3^0 LS-7.5^0 LS (Fig 1). This river is also the boundary of Sidoarjo and Pasuruan.

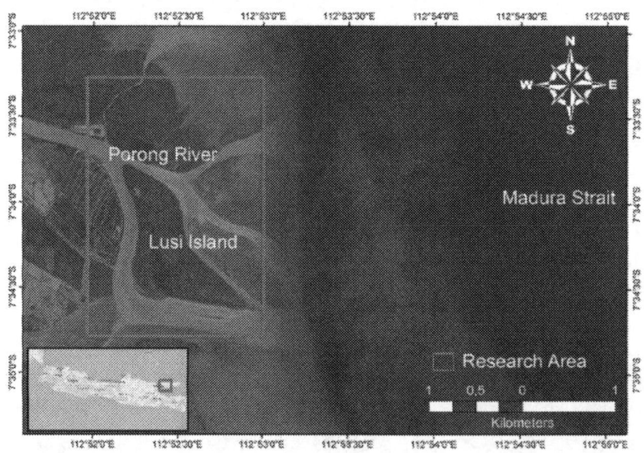

Fig. 1. Study area of Porong River

2.2 Tidal data processing and bathymetri

Tidal data is used as correction to the depth. The depth measurement data using echosounder is corrected by sea level position (MSL, Z_0, and TWLt) at the time of measurement and correction of transducer distance to obtain the actual depth of the water [10].

Tidal has a function to correction of ocean depth data to obtain the true depth of the sea. Taking the data of ocean depth in Muara Porong is done using echosounder assisted by the motor boat with speed of boat speed 5 knot. The data obtained in the form of coordinates along with the depth captured by echosounder sensors. The depth measurement data using echosounder is then corrected to sea level position (MSL, Z0, and TWLt) at the time of measuement and correction of transducer distance to obtain the actual depth of the water. The reduction value (correction) measurement is obtained the following Equation (1) [10] :

$$rt = TWLt - (MSL + Z0) \tag{1}$$

where

rt	:	The amount of reduction (correction) given to the depth measuerement results at time t
TWLt	:	The true sea level position at time t
MSL	:	Mean sea level
Z0	:	The depth of the face below MSL

The result of Equation (4) is further used to obtain the actual depth by using the following Equation (2) [10] :

$$D = dT - rt \tag{2}$$

where

D	:	true depth
dT	:	Transparent depth of the transducer
rt	:	Reduction (correction) tides

Data of sea depth which have been calculated using Equation (1) and (2) got real depth data.

2.3 Current data processing

The calculation to determine the tidal current velocity is performed using the Equation (3) [11].

$$Speed = \sqrt{u^2 + v^2} \tag{3}$$

Likewise to determine the direction of the current, done with the calculation α first by Equation (4) [11].

$$\alpha = arctan\left(\frac{v}{u}\right) \; atau \; \alpha = arctan\left(\frac{u}{v}\right) \tag{4}$$

The calculation to determine the tidal current velocity is performed using Equation 5 [11].

$$Speed = \sqrt{u^2 + v^2} \tag{5}$$

Likewise to determine the direction of the current, done with the calculation α first by Equation (6) [11].

$$\alpha = arctan\left(\frac{v}{u}\right) \ atau\ \alpha = arctan\left(\frac{u}{v}\right) \quad (6)$$

where

u	:	Northern-Southern component of tidal current
v	:	Eastern-Western component of tidal current
u_1	:	Average Northern-Southern component
v_1	:	Average Eastern -Western component

2.4 River flow discharge

River flow discharge (Q), is the amount of water flowing through the cross section of the river per one unit of time, which is usually expressed in cubic meters per second [12]. River debit data is obtained from the Implementing Unit of Irrigation Area Technique, Porong.

2.5 Wind

Wind data obtained from the Meteorology, Climatology, and Geophysics Agency (BMKG) Surabaya. Wind data for 10 years is processed to obtain wind rose and used for wave forecasting.
Stages in wave forecasting:
➢ Correction of location effects [13]:
The correction of the location effect is required because the wind data is derived from the ground station rather than measured directly above sea level or at the seashore. The relationship between sea breeze and terrestrial winds is shown by Equation 7.

$$RL = U_W/U_L \quad (7)$$

where

U_L	=	wind speed on land
Uw	=	wind speed at sea
R_L	=	relationship between the wind speed at sea and the wind on land

➢ Conversion to wind voltage factor [13]
After the correction of the location effect then the next step is to convert the wind speed into a wind voltage factor as shown by Equation 8.

$$U_A = 0.71\ U^{1.23} \quad (8)$$

where

U_A	=	Wind stress factor (m/s)
U	=	Wind speed (m/s)

If the fetch length (F), wind stress factor (UA) and duration are known then high and significant wave periods can be calculated. To estimate the wave height to note the duration of kecepaan wind at sea, wind stress factors (UA) and the effective fetch. This estimation applies only to significant wave that is the average of 33% of the highest wave, of all waves that occur in waters that have been suspected wave height [12].

An effective fetch is a wave generating area with constant direction and wind. The effective fetch equation is denoted by Equation 9.

$$F_{eff} = \frac{\sum X_i \cos\alpha}{\sum \cos\alpha} \quad (9)$$

Where F_{eff} is Effective fecth average, X_i is Length of the fetch segment measured from the wave observation point to the end of the fetch and α is deviations on both sides of the wind direction, using an increment of 6° to an angle of 42° on both sides of the wind direction

2.6 Model of Simulation

The simulation uses a 2-dimensional model that uses averaging to depth, and the parameters used are adjusted for application into the model.

2.7 Hydrodynamic Model

The purpose of simulating hydrodynamics is to obtain a current pattern that will affect the spread of the mud. The builder equation for hydrodynamics is:

The continuity equation

$$\frac{\partial \zeta}{\partial t} + \frac{\partial u}{\partial x} + \frac{\partial v}{\partial y} = 0 \quad (10)$$

The equation of momentum in the direction of x and y

$$\frac{\partial u}{\partial t} + u\frac{\partial u}{\partial x} + v\frac{\partial u}{\partial y} = -\frac{1}{\rho}\frac{\partial p}{\partial x} + 2\Omega\, v\, Sin\,\varphi + F_x \quad (11)$$

$$\frac{\partial v}{\partial t} + u\frac{\partial v}{\partial x} + v\frac{\partial v}{\partial y} = -\frac{1}{\rho}\frac{\partial p}{\partial y} - 2\Omega\, u\, Sin\,\varphi + F_y \quad (12)$$

where ζ is the water level elevation, x and y are Cartesian coordinates, u and v are velocities in the directions $-x$ and $-y$, F_x and F_y are forces in the $-x$ and $-y$ directions, p is pressure, ρ density, g is acceleration of gravity, and φ is the latitude of the Coriolis force.

3 Result

3.1 Depth at Estuary of Porong River

Bathymetry data was taken on March 6-9, 2017. The results of the depth data in estuary of Porong river can be seen in Figure 3. The research area in estuary of Porong river has a depth of 0.1-1.6 m. The average water depth along the river up to the estuary is 0.3-0.5 m indicated by light green color. In the estuary area there is a depth variation, although it is still dominated by green but has been mixed by the blue color indicating the addition of depth. The depth at the estuary ranges from 0.8 to 1.3 m indicated by the blue gradation. The depth of more than 1.4 m indicated by the purple color is on the east side of Lusi Island.

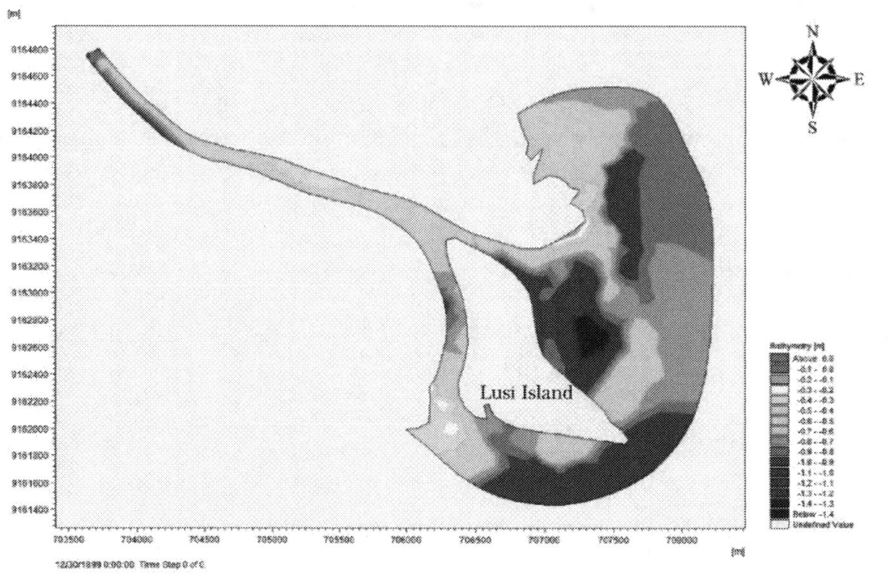

Fig. 2. Bathymetri in Estuary of Porong River, Sidoarjo.

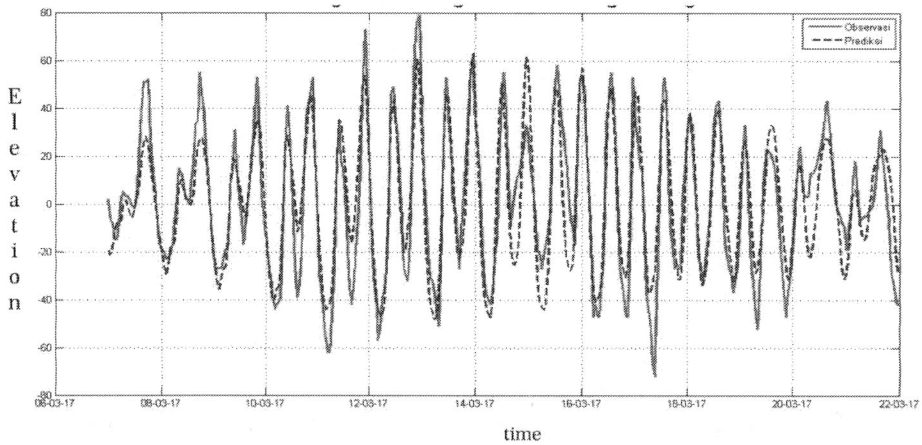

Fig. 3. Tidal charts in Estuary of Porong River.

3.2 Tidal in the Estuary of Porong River

Tidal observations in Porong River were conducted in March 2017 to represent the West season. Tidal types can be determined through tidal observation results from Figure 3 and Table 1. The result shows that the value of Formzahl is 0.62. This figure indicates that the tidal type in Muara Kali Porong includes mixed type tends to double daily, ie within 24 hours of high water position and irregular low water inclined to double daily. This type of tidal is also similar to the results of research from Kisnarti and Prasita [8] and Kisnarti [9]. Mean Sea Level value in Muara Kali Porong is 117 cm from 0 palm.

Table 1. Amplitude and Phase Tidal in Estuary of Porong River

Constants	Frequency	Amplitude	Phase
O_1	0.0387307	11.0247	233.18
K_1	0.0417807	16.2428	329.58
M_2	0.0805114	26.7177	306.79
S_2	0.0833333	17.2121	0.15
M_3	0.1207671	5.0622	209.73
SK_3	0.1251141	2.4077	269.73
M_4	0.1610228	2.4213	233.69
MS_4	0.1638447	4.2672	290.61
S_4	0.1666667	1.6330	313.85
$2MK_5$	0.2028035	1.1038	186.94
$2SK_5$	0.2084474	0.7351	353.03
M_6	0.2415342	0.2516	300.10
$2MS_6$	0.2443561	1.0064	194.35
$2SM_6$	0.2471781	1.6031	180.84
$3MK_7$	0.2833149	0.6011	130.45
M_8	0.3220456	0.2804	178.31

3.3 Tidal in the Estuary of Porong River

The observation of tidal current is done simultaneously with tidal observation in March 2017. The result of tidal current observation can be seen in Figure 4 and Figure 5. The current flow along the river before reaching the estuary is faster than the current in the estuary taht is 1-1.3 m/s markedby gradations of green, yellow, and orange. The current velocity going to the eatuary is at the turn of Lusi Island on the East side hand had increased the current velocity up to 1.6 m/s indicated by the red color, then the current flows towards the estuary towards the East and North waters.

On the west side of Lusi Island from the river to the estuary also experienced an increse in current velocity shown by the red color reaches 1.5 m/s. The curret that passes though the west side of Lusi Island out towards the estuary to the south of the waters. The occurrence of increased speed on the East and West side of Lusi Island when going out the estuary can be caused by physical condition or topography of the waters. Since water flows through the two branches of the creek it causes the current flow to turn and its speed increases. In addition, it can also be caused by the condition of the witdh of the river that originally narrowed then widened when going out towards the estuary.

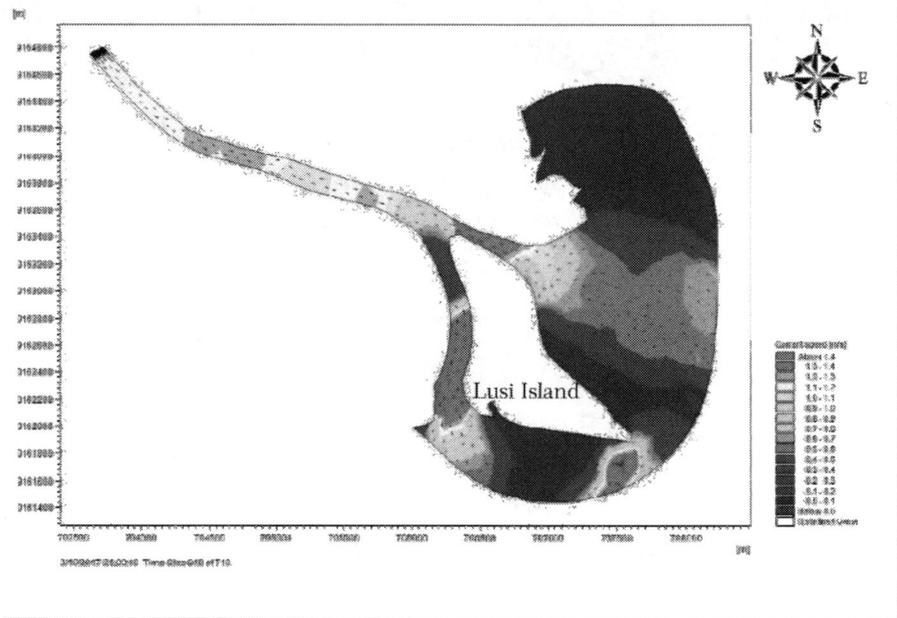

Fig. 4. Current during tide in Estuary of Porong River, Sidoarjo.

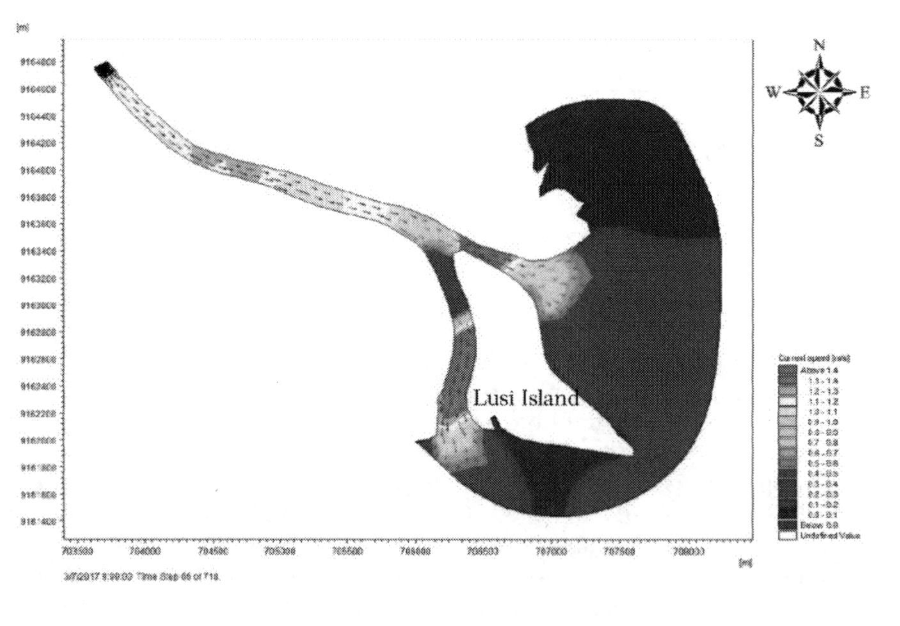

Fig. 5. The current at low tide in Estuary of Porong River, Sidoarjo.

3.4 Discharge in the Estuary of Porong River

Based on the results of river flow data processing in estuary of Porong River during the observation of 462,9 m^3/s, and river flow in March was equal to 569 m^3/s. The value of discharge during the rainy season was not must different from the discharge value during

March 2017 which amounted to 578.9 m³/s. Therefore, the speed of the Porong river flow in March 2017 is more dominated by river flow.

3.5 Wind in the Estuary of Porong River

The main wind in the West season tends to blow from the southwest and the second is from the west (Figure 6a). Speed 1-2 m / s has the highest frequency distribution percentage that is 42.34 % with highest percentage from Southwest direction 10.91 % and seconly from west direction 8.91%. Speed 0-1 m / s has the second highest frequency distribution percentage that is 40.35 % with highest percentage of southwest direction 6.92 % and second from south direction 5.61%. The total data has an average speed of 1.26 m/s and a significant wind speed of 2.29 m/s.

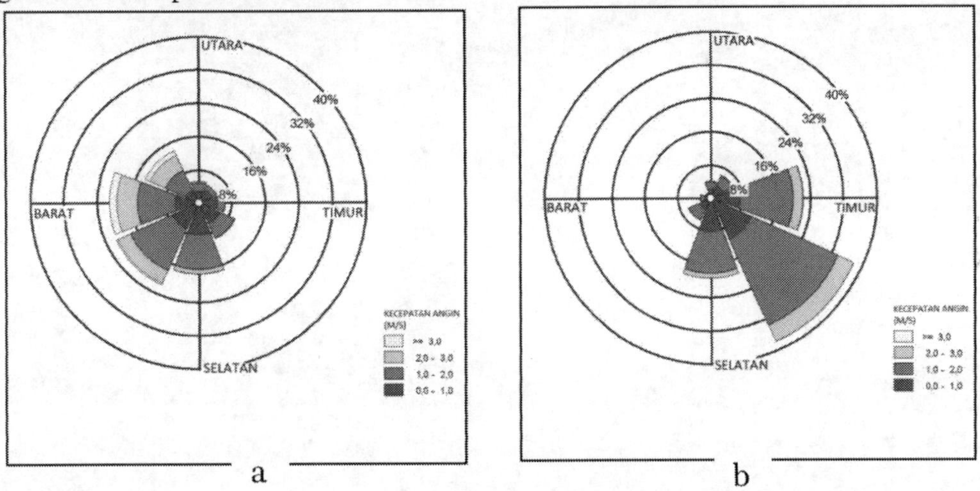

Fig. 6. Wind rose distribution in 2006-2015: (a) western session (b) transition session

The result of the wind data analysis during the transition season I shows that the main winds tend to be from the Southeast and the second direction from the East. (Figure 6b.) The main winds tend to be from the Southeast and the second is from the East. The speed of 1-2 m/s has the highest frequency distribution with the overall total of 53 % with the highest percentage of the Southeast 22.52 % and the second of the East 12.44 %. The speed of 0-1 m/s has the second highest frequency distribution with the overall total of 38.35 % with the highest percentage of the Southeast direction of 11.00 % and the second is from the East 7.35 %.

The average wind speed at Western and first transition session are 1.26 and 1.15 m/s and a significant wind speed of 2.29 and 1.78 m/s hence the average wind speed or significant less generating a minimum U_A so that the wind is said to be weak and can not be used to predict the height and significant wave period.

3.6 Suspension Sediment Analysis

Using hydrodynamic model, simulation has been done based on tidal data input, current discharge and wave. The average different drift in the research area in estuary of Porong river for 15 days way 0.2-0.8 kg/m³ as shown in Figure 8 through the dominant blue color gradation. In areas with high current velocities, the suspended sediment is higher at 3 kg/ m^3, indicated by red. Sediment suspension increasingly heading towards the estuary until the estuary is not too big about 0.2-0.8 kg/m^3. In the area leading to the sea the suspension becomes 1.4-2 kg/m^3, this may occur due to the confluence of river and sea water so that the turbulence is greater and causes the sediment to rise (Figure 7).

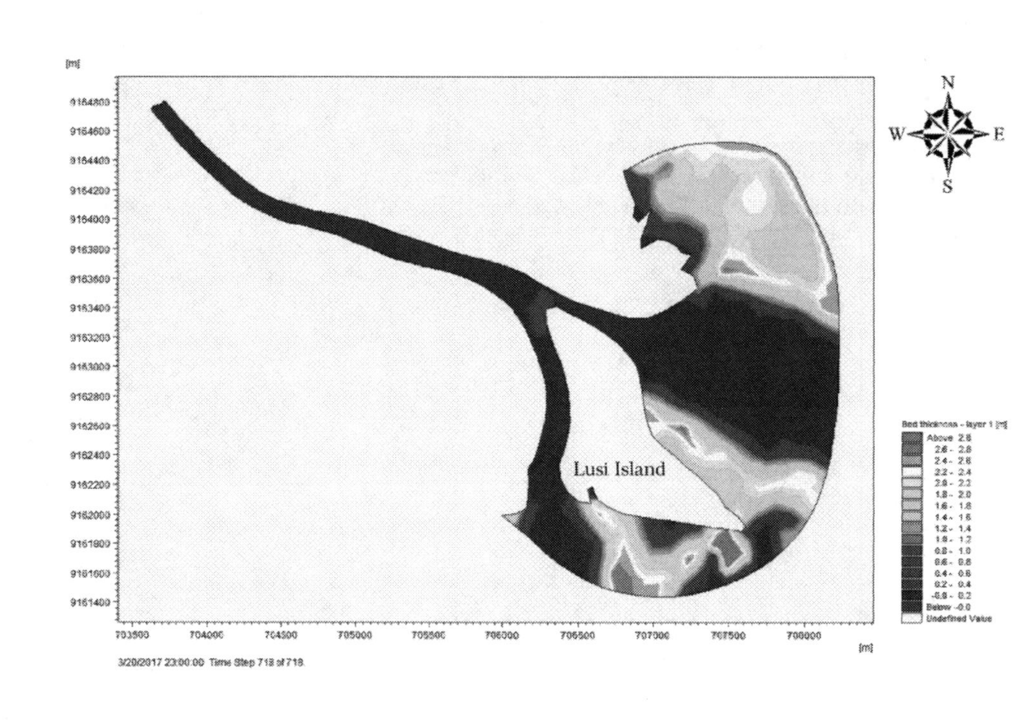

Fig. 7. Sediment suspended in Estuary of Porong River for 15 days.

3.7 Sedimentation

Sediment samples were taken at 3 stations. The result of laboratory analysis about sediment samples showed that at station 1, sand content was more dominant with percentage of 51.14 %, followed by fine grain (silt + clay) 48.68 % and gravel 0.19 %. The sediment composition at stations 2 and 3 differs from station 1. Station 2 and 3 is more dominated by fine grains (silt + clay) with composition of 81.30 % and 67.41 %. The next composition is the sand of 18.56 % and 32.46 %, while the gravel at stations 2 and 3 is only 0.14 %.

The sedimentation process with a 15-day simulation takes place on the north-east of Lusi Island with a thick sediment ranging from 1.6 to 2.6 m with gradations of green to light orange (Figure 8). In addition to the Northeast Lusi Island, sedimentation also occurs on the lower side or Southeast and South Lusi Island shown by green to red. The southern side has a larger sediment thickness than the Southeast side, it can be seen that the red color is more on the South side which shows the figure of 2.8 m. Sediments on the Southeast side average around 2-2.4 m.

Sedimentation occuring on the Northeast and Southeast side can be caused by the flow of currents from the river through the eastern branches. The current from that direction turns northeast to north and southeast and southward. At the moment the current flows towards the southern sediment hampered by the shape of Lusi Island. In addition, the accumulation of sediments is also due to tides, when the tide then sediment will be transported towards the estuary. This may affect althouh the tidal effect on this location is not very dominant.

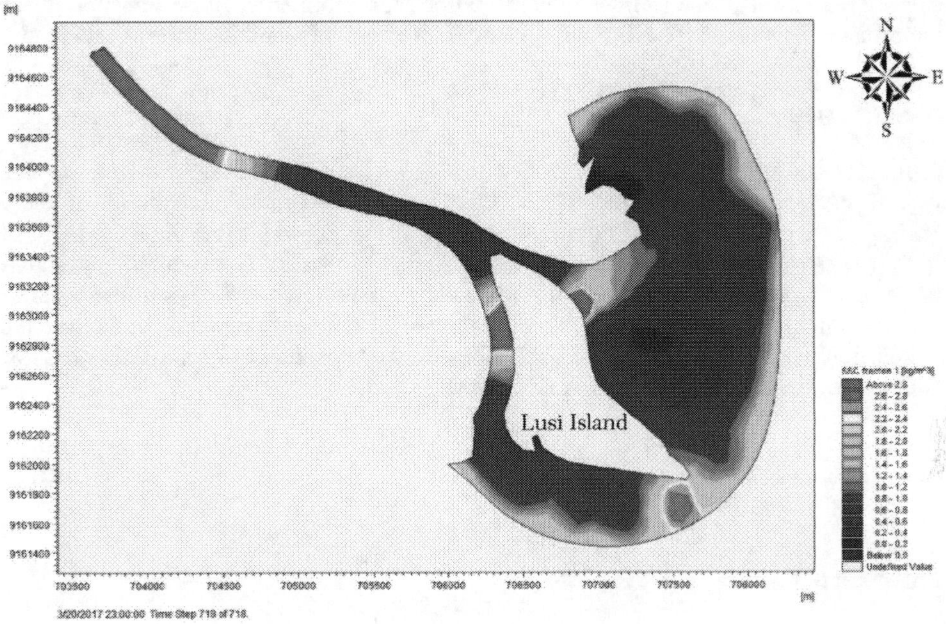

Fig. 8. Simulation of basic sediment in Estuary of Porong River for 15 days.

4 Discussion

The morphological changes of estuary of Porong river are strongly influenced by Oceanographic factors, such as currents and waves. From the results of data processing for the oceanographic parameters, it can be concluded that the influence of the change of estuary Porong most dominant in the wet season is caused by non tidal currents. Non-tidal currents are residual currents with varying effects. To find out the dominant cause of non tidal currents is done the data processing of river flow and wind data as input from wave. From the river flow data processing known the flow velocity of river discharge with average flow velocity of 578,9 m^3/s. Pahlevi et al. [14] also atated that the source of sediment material in the coastal area area is caused by Sidoarjo Coastal. There are many river estuaries that have relatively high river flow (500 m^3/s) plus since 2006 the abundance of sediment comes from Lapindo mudflow ehich is flowed into the Strait of Madura through Porong river. Wind data with a significant 10-year wind speed of 2.22 m/s and an average speed of 1.41 m/s is very weak to generate wave energy. From the analysis of river flow rate velocity and wave forecasting, river flow has the greatest influence on the change of the Porong River Estuary as evidenced by the current separation that is dominated by non tidal currents rather than tidal currents. River estuaries dominated by river discharges with relatively small waves will carry large amounts of upstream sediment transport.

Hydrodynamic model simulation shows that sediment suspension progresses toward the estuary until the estuary is not roo big, otherwise the sediment suspension that goes to the sea increases. This can happen because of the confluence of river and sea water so that the turbulence is greater and causes the sediment to rise.

Sedimentation occuring on the notheast and souteast side can be caused by the flow of currents from the river through the eastern branches. The current from that direction turns northeast to north and southeast and southward. At the moment the current flows towards the southern sedimen hamoered by the shape of Lusi Island. In addition, the accumulation of sediments is also due to tides, when the tide then sediment will be transported towards the estuary. This may affect although the tidal effect on this location is not very dominant

5 Summary

From this research, it can be concluded that (1) wind velocity is very weak to generate wave energy in the research location so that less give maximum contribution in change of Porong river estuary; (2) the dominant current at the Porong River Estuary is a nontidal flow; (3) estuary of Porong River can be classified as estuary type which is dominated by river discharge; (4) sedimentation process with a simulation of 15 days in the West Season occurred in the Northeast of Lusi Island with a thick sediment ranged from 1.6 to 2.6 m; (5) the current results are very precise and the estuary of Porong river during the West Season is more dominated by the very high river flow.

We thank the Indonesian institutions that have contributed significant support and valuable information to the research: Ministry of Research, Technology and Higher Education, Republic of Indonesia and Hang Tuah University, Surabaya, Indonesia.

References

1. A. Harnanto. Peranan Kali Porong dalam mengalirkan lumpur Sidoarjo ke laut, Badan Pelaksana Badan Penanggulangan Lumpur Sidoarjo (BAPEL-BPLS) (2011).
2. T. C. Jennerjahn, I. Jänen, C. Propp, S. Adi, S. P. Mugroho. Environmental impact of mud volcano inputs on the anthropogenically altered Porong River and Madura Strait coastal waters, Java, Indonesia. *Estuarine, Coastal and Shelf Science* **130**, 152-160 (2013)
3. N. Maryantika, C. Lin. Exploring changes of land use and mangrove distribution in the economic area of Sidoarjo District, East Java using multi-temporal Landsat images. *Information Processing in Agriculture* **4**, 321-332 (2017)
4. A. S. Riyadi, B. Sarwono, Sudiwaluyo. Permodelan aliran sediment di Muara Kali Porong, *Seminar Nasional Aplikasi Teknologi Prasarana Wilayah*. ISBN 978-979-18342-1-6 (2009).
5. H. Bachtiar, N. Fronto, R Fitri. Simple model of two dimensional sediment movement in Porong River. *Journal of Marine Geological Institute* **9** (**3**). ISSN 1693-4415 (2011).
6. W. Atmojo. Studi penyebaran sedimen tersuspensi di Muara Sungai Porong Kabupaten Pasuruan, *Buletin Oseanografi Marina Oktober* **1,** ISSN 2089-3507, 60-81 (2011).
7. U. Hernawan, K. Budiono. Karakteristik dan distribusi lumpur Sidoarjo sepanjang Sungai, Estuari dan Perairan Porong. *Jurnal Geologi Kelautan*. **11(2)**, 91-99 (2013)
8. E. A. Kisnarti, V. D. Prasita, Perubahan morfologi di Muara Sungai Porong, Sidoarjo. *Pertemuan Ilmiah Nasional Tahunan X ISOI.* ISBN. 978-602-18153-2-8, 381-387 (2014).
9. E. A. Kisnarti, Pasang surut dan arus pasang surut di sekitar Muara Sungai Porong Sidoarjo. *Jurnal Akuatika Indonesia.* **1** (**2**). ISSN. 2528-052X, 101-108 (2016).

10. A. D. Saputra, H. Setiyono, A. Anugroho, Pemetaan batimetri dan sedimen dasar di Perairan Karangsong, Kabupaten Indramayu, Jawa Barat. *Buletin Oseanografi Marina* **5**(**1**). 38 – 43 (2016).
11. C. M. Simatupang, H. Surbakti, A. Agussalim, Analisis data arus di Perairan Muara Sungai Banyuasin Provinsi Sumatera Selatan. *Maspari Journal*, **8** (**1**), 15-24 (2016).
12. B. Triatmodjo, *Teknik Pantai*. Beta Ofset. Yogyakarta (1999).
13. Coastal Engineering Research Center. 1984. *Shore Protection Manual*. Departemen of the Army. Washington DC.
14. A.M. Pahlevi, Wiweka, Analisa sedimentasi di Muara Kali Porong akibat pembuangan lumpur Lapindo menggunakan data citra satelit Aster. Jurnal Ilmiah Geomatika, **16**(**2**), 15-24 (2010).

Sea level rise impact modelling on small islands: case study gili raja island of east java

Zainul Hidayah[1,*], *Luhur Mukti* Prayogo[1] , and *Maulinna Kusumo* Wardhani,[1,2]

[1]Department of Marine Science, Trunojoyo University of Madura, East Java Indonesia
[2]Post Graduate Student, Faculty of Marine Technology, Sepuluh Nopember Institute of Technology, East Java Indonesia

Abstract. Coastal regions and small islands are areas that will be adversely affected by the phenomenon of sea level rise globally. In general, Sea Level Rise (SLR) will result in coastal impacts as follows: increased frequency and intensity of floods, changes in ocean currents and widespread intrusion of sea water. This research was conducted in Gili Raja Island of Sumenep Madura. Objectives of this research were to demonstrate the ability of combining remote sensing and GIS method to determine the impact of SLR on a small island and to model its scale using different scenario. GIS based run-up model were performed to estimate and predict the impact of SLR to the island's area. Three water level scenario (0.5 m, 1.0 m and 1.5 m) were applied. The results showed that in the first scenario 8.73% of the island was flooded by sea water, furthermore in two other scenario the flooded area was increase significantly (15.88% and 22.38%). Keywords : sea level rise, small island, geographic information system

1 Introduction

Sea Level Rise (SLR) is one of the important issues that must be faced by coastal or island countries in the world. Some studies explain that at least 3 factors are considered to be the main cause of SLR. Those factors are: melting ice in both earth's poles, extreme climatic events and land subsidence. According to the IPCC (International Panel on Climate Change) report, the average global surface temperature increased from 0.3 to 0.6 °C since the late 19th century. Moreover, until 2100 the earth's temperature is expected to rise around 1.4 – 5.8 °C [1]. The global rise of Sea Surface Temperature (SST) lead to melting ice in the North and South Pole of the Earth. As the consequence, there was an increase in sea level. It is estimated from the year 1999 - 2100 sea level could rise around 1.4 - 5.8 meters [2]. Indonesia was predicted to suffer a significant loss of coastal land as the result of SLR's impact. Using numerical modeling, it was estimated that in 2050 over 30,120 km^2 of coastal land will be inundated. Moreover, in 2100 the flooded areas were estimated to increase approximately 90,260 km^2 [3].

The above figures emphasized this natural event needs to be taken into account in all coastal area management activities, as it can have a direct impact on coastline retreat and may disrupt the population's assets and economic development of the region and even lead to

[*] Corresponding author: zainulhidayah@trunojoyo.ac.id

displacement of populations that inhabit vulnerable areas along the coast [4]. In addition to the SLR threat, several countries in the Pacific have reported the lost of small island areas caused by sea level rise. Therefore, a global awareness should be raised among the scientists and governments to predict the occurrence and impact of SLR events in coastal areas particularly small islands.

Coastal regions are areas with great potential for economic development, but very vulnerable to sea level rise. Many large industries are established in coastal areas because of the transportation access. Developments in coastal areas are then encouraging the growth of the economy in the region. The value of economic losses will be very different for every type land use. The magnitude of the value of economic losses will depend on the level of productivity of the flooded land. The loss of land due to sea level rise, not only cause economic losses but also loss of biodiversity in the region [5].

Sea water run-up model is one way to estimate the impact of Sea Level Rise on coastal areas in the future. Using combination of remote sensing data and Geographic Information System (GIS) methods, distribution and impact of sea level rise can be determined [6]. This research is conducted in Gili Raja Island of Sumenep Region Madura. Some parts of the island have been inundated by sea water during the tide. This phenomenon was suggested to be an indicator of SLR effect on the island. Objectives of this research are to demonstrate the ability of combining remote sensing and GIS method to determine the impact of SLR on a small island and to model its scale using different scenario.

2 Material and Method

Coastal impact models can help determine the vulnerability of areas and populations to changes in sea level. Model outputs may be used to guide decisions about the location and design of future protected areas and development, and to prioritize adaptation of existing protected area investments [7]. A number of studies have been done to develop models of SLR impact on coastal areas. Most of the models were developed using numerical and empirical approach [8,9]. Furthermore, some of the studies incorporate SST and SLR to create Cyclone Prediction Model [10], whereas this research focuses on using GIS model to generate spatial information of SLR impact to a small island. GIS modeling approaches have been widely used to provide spatial information of coastal regions as well as to model the effect of natural and anthropogenic events to the environment. However, specific requirements of data-set have become issues that determine the reliability and validity of the model. This research employed common and generally available spatial data such as land use types and elevation. Land use map of the island was built based on ALOS-AVNIR-2 satellite imagery data.

Land use analysis was performed using Maximum Likelihood Algorithm (MLA) continued by ROI statistical calculation to determine the Separability Index of each land use types. The main assumption of MLA is that the spectral response for each class is normally distributed [10]. Moreover, for an *(m)* class case, the decision process calculates the probability of a pixel belonging to each of *(m)* predetermined classes, and the class with the highest probability is assigned to that pixel [11]. The estimated probability density function or Likelihood function for a given class *(wi)* is calculated as :

$$p(x; wi) = \frac{1}{(2\pi)^{\frac{1}{2}}} \exp\left[-\frac{1}{2}\frac{(x-\mu i)^2}{\sigma i^2}\right] \quad (1)$$

where, x is one of the brightness values on the x-axis, ˆμ is the estimated mean of all the values in the class and σi is the estimated variance of all the measurements in this class. In addition, for multispectral satellite imagery data which usually consists of more than one band for a class of interest, assuming the number of bands is (*n*), then the n-dimensional multivariate normal density function can be calculated as follows:

$$p(X; wi) = \frac{1}{(2\pi)^{\frac{n}{2}}} \exp\left[\frac{1}{2}(X - Mi)^T Vi^{-1} (X - Mi)\right] \quad (2)$$

where, (M)is the mean vector for class i,(V)is the covariance matrix for class i, |Vi |is the determinant of the covariance matrix for class i, Vi^{-1} is the inverse of the covariance matrix for class i, and $(X - M)^T$ is the transpose of the vector (X-M).

Afterwards, land use types were used to generate Roughness Surface Map [10] as it can be seen in Table 1. Another important data for this analysis was a Digital Elevation Model of island's surface. This data was obtained from 30x30 meter SRTM images. This image was used to generate a surface contour and slope map which very useful to delineate the boundary of sea water inundation. Furthermore, analysis was continued by modeling of water inundation height (run-up) using mathematical equation as follows [11].

$$H_{LOSS} = (167n^2/H_o^{1/2}) + 5 \sin S \quad (3)$$

Explanation : H_{LOSS} : decrease of water level from original distance
n : roughness index
H_o : water level at coastline
S : Slope

Table 1. Roughness Surface Coefficients

No	Types of Land Use	Roughness Coefficients Index
1	Settlements (buildings and houses)	0.080
2	Mixed Vegetations	0.070
3	Mangrove Vegetations	0.025
4	Wasteland	0.015
5	Fish/Salt Ponds	0.010

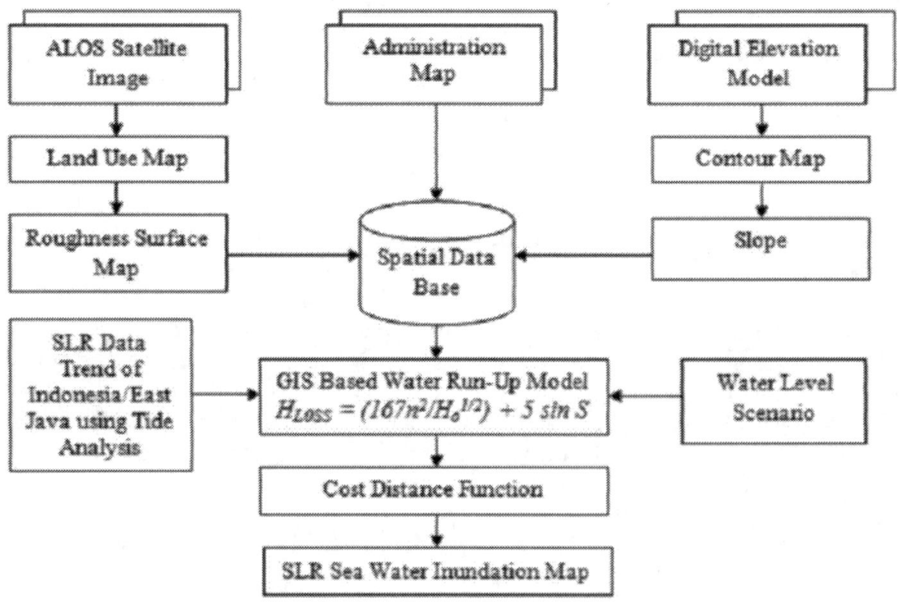

Fig. 1. Data Analysis Flow Chart

3 Result and Discussion

Gili Raja Island is located in the Madura Strait, south of Sumenep district. Geographic coordinate of the island is 07⁰13'10.774"S and 113⁰46'40.345"E. Total area of the island approximately 1,139 Ha, it consists of 4 villages with total population around 15,000 people. Most of the people are fishermen and farmers. In addition, access to the island can be reached from Talango Island which is 11.93 miles away, whereas the nearest island is Gili Genting about 5.30 nautical miles east of the island. The island can be classified as a flat island, elevation ranges from 0-50 meters above sea level. Sandy beaches can be found on the north and east part of the island. Meanwhile, the south and west parts are dominated by rocky beaches. Moreover, traditional salt ponds were also found on the island, located in the north and operated by local communities. Residential areas are distributed on entire island, but mostly concentrated in the south.

Table 2. Land Use Classes of Gili Raja Island

No	Types of Land Use	RCI	Total Area (Ha)
1	Settlements (buildings and houses)	0.080	178.25
2	Mixed Vegetations	0.070	232.58
3	Mangrove Vegetations	0.025	6.49
4	Empty Lands	0.015	439.88
5	Fish/Salt Ponds	0.010	281.79

ALOS satellite imagery data in this study was used to identify land use types on island's surface. Supervised classification method with MLA was employed to group pixels based on their spectral value compared to selected sample areas. As the results, land use types in this island were divided into 5 classes ie settlements (houses and buildings), mixed vegetation cover, mangroves, fish ponds and bare land. Satellite image analysis and land use types calculation showed that most of the Gili Raja island area was dominated by wasteland (38.62%), fish ponds (24.74%), mixed vegetation cover (20.42%), settlements (15.65%) and mangroves (0.57%). The results of overall accuracy was 78.54%, it considered to be acceptable for this study (OA_{value} > 75%). Moreover, Kappa value=0.702 represents probability of 70.2% better accuracy than if the classification resulted from random unsupervised classification instead of using MLA. Table 2 showed the complete results of land use classification of Gili Raja Island. Afterwards, land use classes were converted into roughness surface map based on roughness coefficient index.

Table 3. MLA Results of Gili Raja Land Use Classification

Class Name	RT	CT	NC	PA (%)	UA (%)	*Kappa* Value
Settlements	102	60	59	58.84	94.17	0.812
Mixed Vegetations	135	75	65	57.85	77.17	0.739
Mangrove Vegetations	25	20	18	78.15	85.29	0.837
Wasteland	55	48	46	87.27	82.58	0.624
Fish/Salt Ponds	10	8	9	77.12	75.47	0.792
Overall Classification Accuracy : 78.54% Overall Kappa Value : 0.702						

RT : Reference Total; CT : Classified Total; NC : Number Correct
PA : Producer's Accuracy; UA : User's Accuracy

Beside land use types of the island, elevation and surface's slope data are important input to build SLR inundation model. Those data was derived from Digital Elevation Model (DEM) with 30x30 pixel resolution. Analysis results showed that 63.22% of the islands was dominated by low land and flat areas (elevation ranges from 0-17 meters above sea level and slope < 15%). These areas were mostly located in the north side of the island. Whereas higher and steeper area of the island were located in the south side where elevation ranges from 18-50 meters and slope > 25%. According to this condition, therefore settlements, cultivated areas and fish/salt ponds were concentrated in the north.

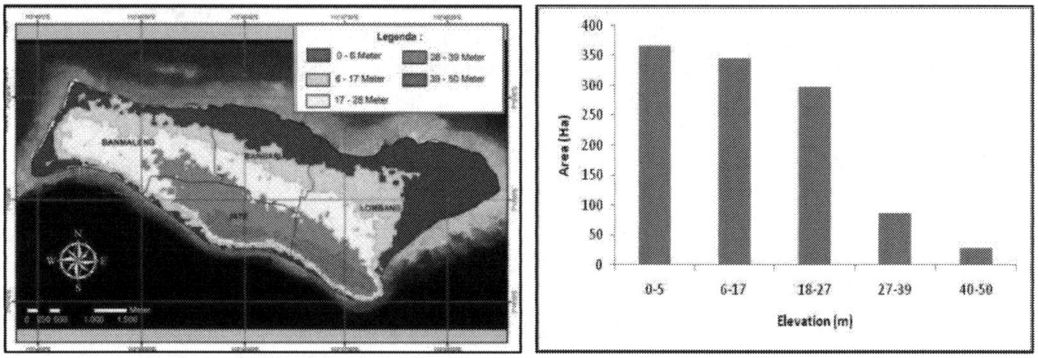

Fig. 2. Digital Elevation Model (DEM) of Gili Raja Island

Beside land use types of the island, elevation and surface's slope data are important input to build SLR inundation model (Figure 2). Those data was derived from Digital Elevation Model (DEM) with 30x30 pixel resolution. Analysis results showed that 63.22% of the islands was dominated by low land and flat areas (elevation ranges from 0-17 meters above sea level and slope < 15%). These areas were mostly located in the north side of the island. Whereas higher and steeper area of the island were located in the south side where elevation ranges from 18-50 meters and slope > 25%. According to this condition, therefore settlements, cultivated areas and fish/salt ponds were concentrated in the north.

Furthermore, tide analysis results explain that in the past 10 years there have been gradual increases in water level. In 2006 highest water level during high tide was approximately 1.08 meter, however in 2016 the level was increased and reached around 1.20 meter. It means that in the last 10 years, water level nearby Gili Raja Island has increased around 0.016 meter/year. This result can be used to predict the height of water level in the next 50 years or so. Afterwards, using GIS-based water run-up model combined with cost distance function, flooded area of the island can be estimated at the same period. The prediction results are explained in this following table.

Table 4. Modelling Results

Scenario	Year	Highest Tide (m)	Increasing Water Level (m)	Flooded Area (Ha)	% Area
1	2066	2.04	0.8	99.45	8.73
2	2116	2.84	1.6	180.89	15.88
3	2166	3.64	2.4	254.89	22.48

As can be seen from the results, predicted inundation on Gili Raja Island is started from the north and west side of the island (Figure 3). The higher sea level rise, the more area will be flooded. This happens because those parts of the island are in the lowlands area. The flood can easily reach a long distance from beach line when it comes to flat area. In addition, as the location has become center of the population settlement, it is estimated that sea level rise will have a serious impact on island's population. Further analysis by overlaying inundated areas and land use map allows an assessment of the impact of inundation. The results are shown in Table 5.

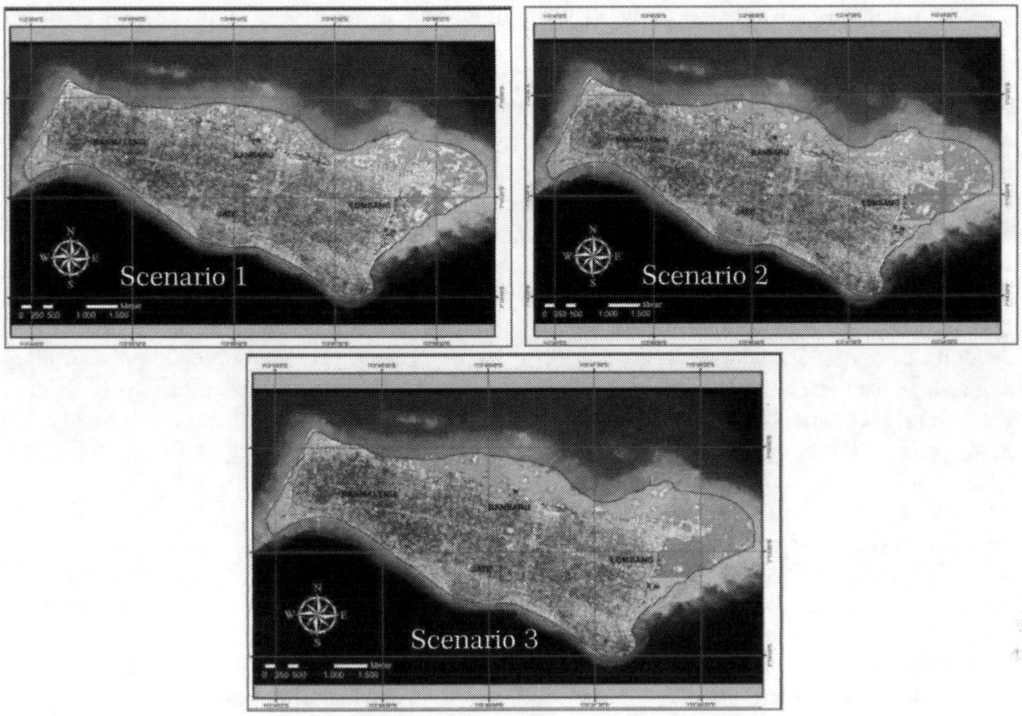

Fig. 3. GIS-Based Run Up Modelling of Gili Raja Island

Table 5. Affected Land Use Types

Land Use Types Scenario/ Area (Ha)	1	2	3
Settlements			44.85
Mixed Vegetation	33.75	67.97	78.58
Mangrove	1.98	4.37	5.78
Wasteland	17.83	42.14	45.28
Fish/Salt Ponds	45.89	66.41	80.4
Total Area (Ha)	99.45	180.89	254.89

Based on scenario 1, the most vulnerable area are fish/salt ponds and mixed vegetation along the coastline followed by wasteland and mangroves. In total, inundation will cover around 99.45 Ha (8.73%) of the entire island's area. It is estimated that inundation will obviously damage the coastal environment. Furthermore, scenario 2 predicts more flooded area on the same land use type as the previous one. Inundated areas are increase significantly into 180.89% (15.88%). Finally, the same trend continues to happen in the last scenario, where up to 254.89 Ha (22.7%) areas are flooded. It brings damage not only to environment but also for settlements. GIS-based run up modeling explain that SLR on small island will result in flooding of low-lying areas, increased sea water intrusion, coastal erosion, decline in fisheries resources and the loss of island's biodiversity.

Climate change is projected to affect small islands, partly by exacerbating natural disasters and partly through more gradual effects such as rising sea level. The characteristics of small islands make them vulnerable to SLR and other natural disasters due to climate change and global warming. In contrast to larger islands or main lands, natural disaster on a small island can lead to a fatal disruption of economic processes, serious impact on environmental and extensive damage to settlements [13]. Permanent flooding from SLR is different than the temporary flooding from storm surge or precipitation. Intermittent flooding recedes once a storm passes while SLR flooding is permanent and can be expected to encroach further inland over time. Sea level rise not only permanently alters the coast line; it also widens the area vulnerable to storm surge [14]. In addition, as the consequence of SLR, water shortages and the shrinking of land suitable for agriculture and fish ponds would cause other social and political disruptions, including forced migration and conflict. Small island will thus face much larger economic costs from climate change than larger peers. The impact on important economic sectors (agriculture, tourism and fishing) and pressures on ecosystems could exacerbate poverty and emigration [15].

Undoubtedly, SLR might cause severe impact on environment and human lives on coastal and small island areas. Nonetheless, by having prediction and estimation on potential impact that could occur in the future, preventive and adaptation actions may possibly be taken into account. Local governments have responsibility to design and provide domestic policies that can reduce the direct human and economic costs of natural disasters particularly SLR. There are several recommendations that could be done to protect small island from SLR. Firstly, is engineering solutions by continually pumping more sand onto beaches or building higher berms and sea walls around communities and infrastructure. Since building solid construction surrounding the island will need a large amount of money and its effectiveness is still questionable, thus this solution seems unlikely to be implemented in Gili Raja Island. The second solution is to protect the island by restoring mangrove ecosystems. Only a small portion of Gili Raja Island is covered by mangrove vegetations. Mangrove forest ecosystem acts as a natural barrier against potential natural disaster such as storm, SLR and tsunami. Moreover, recent studies revealed the ability of mangrove ecosystems to cope with SLR. It is interesting to note that the height of the soil surface in mangrove areas is often dynamic, and in some cases appears to be building up at rates of between one and 10 millimetres every year. Meanwhile the global mean SLR is currently 3mm per year, meaning that many mangrove areas build up soil at a rate which keeps pace with the sea. The main reason behind it is the ability of mangroves to trap sediment as it is carried by rivers, and the work of their roots beneath the surface [16]. Compare to the previous solution, restoring mangrove and wetlands needs less money, however it takes more time in order to gain the results.

4 Conclusions

This study explains the usefulness of GIS modeling for coastal disasters prediction and management. By combining the result of satellite imagery data, DEM and tide analysis, the inundation map can be produced and the future impact of SLR on island's surface can be predicted and calculated. Small island resources and its economic value are important and beneficial to the regions development. Spatial model using GIS for small island inundation mapping is considered as an important task to support the management plan of SLR risk reduction initiatives. It is recommended that to provide protection for the island against SLR, mangrove rehabilitation should be taken into account.

In order to develop more comprehensive studies in the future, several subjects remain to be discussed for instance is the issue related to validity or accuracy assessment of the GIS model. In this study, accuracy assessment was conducted only to the results of image classification. It is recommended for future study to add accuracy assessments on inundation map particularly to compare between existing and the predicted condition. Another issue is the calculation of economic loss due to SLR. This study has demonstrated the ability of GIS to calculate inundated area based on different land use types. It is therefore suggested that economic value and productivity per class of land use should also be taken into account for future studies to gain more comprehensive data analysis in particular economic overview of the SLR impact on small islands.

References

[1] Andre, C.*Analysis of Damages Due to Marine Flooding and Evaluation of Building Costs from Insurance Data*: Perspectives from the Johanna (2008) and Xynthia (2010) Storms PhD Thesis. James Cook University, Australia. (2013)

[2] Berryman, K. *Review of Tsunami Hazard and Risk in New Zealand*. NewZealand: insitute of Geological and Nuclear Sciences. (2006)

[3]. Susandi, A. & Herlianti, I. Dampak Perubahan Iklim terhadap Ketinggian Muka Laut di Wilayah Banjarmasin. Institut Teknologi Bandung. Bandung.(2008)

[4] Diposaptono, S. *Karakteristik Laut pada Kota Pantai*. Direktorat Bina Pesisir. Direktorat Jenderal Pesisir dan Pulau-pulau Kecil. Kementerian Kelautan dan Perikanan. Jakarta. (2012)

[5] Marfai, M.A, dan King, L. *Coastal Flood Management in Semarang, Indonesia*. International Journal of Geography Environment and Geology. Vol **11** No.3 : 89-97.DOI 10.1007/s00254007-1101-3. (2008)

[6] Edhi, F. *Application of Satellite Data to Analyze Inundation Potential and The Impact of Sea Level Rise*. Journal of Remote Sensing. Vol **9** No. 2 : 140-151.(2012)

[7] McLeod, E., Poulter, B.,Hinkel, J., Reyes, E. & Salm, R. *Sea Level Rise Impact Models and Environmental Conservation : A Review of Models and Their Applica*tions. Ocean and Coastal Management. Vol **53** Issue : 9. (2010).

[8] Ali, A. *Climate Change Impacts and Adaptation Assessment in Bangladesh*. Climate Research Vol **12** (2) 109–116. (2000)

[9] Li, Z. & Zhang.J. *Calculation of Field Manning's Roughness Coefficient*. Agricultural Water Management Vol **46**. Elsevier. 153-161. (2001)

[10] Shivakumar, B.R & Rajasakararadhya, S.V. *Spectral Similarity for Evaluating Performance of Traditional Classifiers*. Proceeding. IEEE WISPNET Conference. (2017)

[11] F. S. Erbek., Özkan,C & M. Taberner. *Comparison of Maximum Likelihood Classification Method with Supervised Artificial Neural Network Algorithms For Land Use Activities*", International Journal of Remote Sensing. Vol **25**.No 9. pp. 1733–1748. (2004)

[12] UN OHRLSS. *The Impact Of Climate Change On The Development Prospects Of The Least Developed Countries And Small Island Developing States*. Research Report. (2009)

[13] Fraudenberg, R. *et.al*. Under Water : How Sea Level Rise Threatens the Tri-State Regions. Regional Plan Association. New York. (2016)

[14] Webb, A.P & Kench, P.S. *The Dynamic Response of Reef Islands to Sea Level Rise: Evidence from Multi-Decadal Analysis of Island Change In The Central Pacific*. Global and Planetary Changes. Vol **5**. No 3. (2010)

[16] The Nature Conservancy. 2016. *Mangroves Could Survive Sea Level Rise if Protected*. Available on-line. http://www.cam.ac.uk/research/news/mangroves-could-survive-sea-level-rise-if-protected. Accessed Date : 22 Februari 2018.

Wave energy potency in East Java coast

Heri Saptono Warpindyasmoro[1,*]

[1]Electrical Engineering Department, Petra Christian University, Jl. Siwalankerto 121 - 131 Surabaya, Indonesia

Abstract- It is predicted that in 2025, Indonesia energy needs is going to reach 280% of current needs. Presently, the renewable energy has only contributed 5.7% of total energy supply. The aim of this research is to investigate wave energy potency in several coasts of East Java which were Pacitan, Jember, Besuki and Tuban. Wave data (significant wave heights and wave periods) were obtained from the ERDDAP server (Environmental Research Division's Data Access Program). Southern East Java Coast showed potency for ocean wave energy resources. The coast of Pacitan had energy of 232 MWh/m/year and the coast of Jember had energy of 190 MWh/m/year.

1 Introduction

Energy demand in Indonesia has increased by 6% per year. In 2025, an increase of energy demands is likely to be 180% of present energy supply which means from 3 million BOEP (Barel of Oil Equivalent per Day) to 8.3 million BOEPD. Proportion of renewable energy supply among all energy sources was only 5.7%. Ocean energy has long been studied by researchers both in Indonesia [1, 2] and other countries [3]. The researcher have not only studied the source of ocean energy but also have studied technology of wave energy converter [4,5] A study of wind, tidal, ocean currents and waves in the Sunda Strait as an alternative energy has previously been conducted [6]. It was found that Sunda Strait had waves with maximum wave height 3.2 m and periods 17 second which could produce 92 kW/m energy. Furthermore, tidal energy was not feasible enough because the height was only about 0.5 m to 0.7 m. Effect of wind on ocean wave in South Coast of West Java has also been studied [7]. In this study, it was found that wind and wave patterns varied greatly according to seasonal patterns. Tidal modelling has been conducted in Alas Strait (between Lombok Island and Sumbawa), the estimated energy obtained in one year was 330 GWh [8]. Researches have not only been conducted in Indonesia but also in Canary Islands [9] and Portuguese [10] where the prototype of ocean wave energy generation was made.

* Corresponding author: herisw@petra.ac.id

2 Research Methods.

The power of ocean waves can be calculated as in equation (1) [9, 11]

$$P = \frac{\rho g^2}{64\pi} H_s^2 T_e \tag{1}$$

where P is the energy flux per width of wave (W/m), H_s is the significant wave height (m), T_e is the wave energy period (s), ρ is the density of sea water (kg/m^3) and g is the acceleration of gravity (m/s^2).

According to equation (1), significant wave heights and wave periods data are needed to calculate the power generated by sea waves. These data were obtained and provided by third parties. In this study, the data were obtained from https://coastwatch.pfeg.noaa.gov/erddap/index.html. ERDDAP (the Environmental Research Division's Data Access Program) was a server that provided oceanographic data. The data did not cover all ocean surface. However, it was recorded as points in a grid with 0.5° space of latitude and longitude lines. Moreover, the data used in this study were from January 1, 2013 to December 31, 2013 which ware collected daily in every one hour. Figure 1 shows the location to be studied.

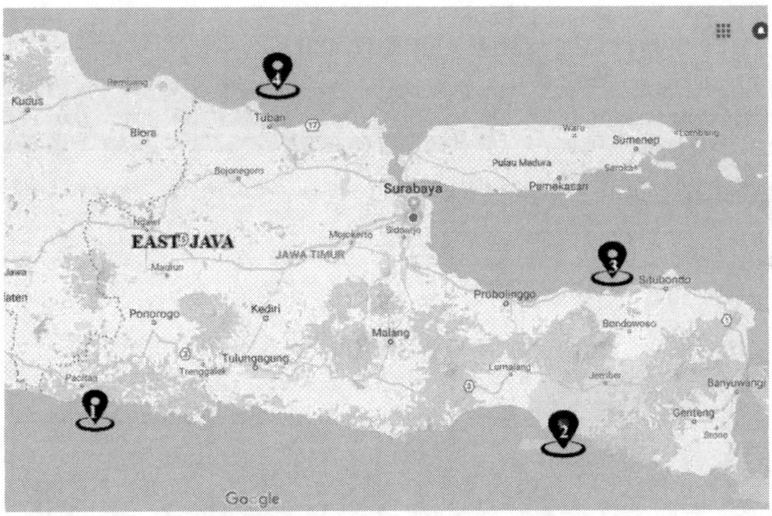

Fig. 1. The location of wave energy to be analysed, namely: Pacitan (-8.5, 111.0), Jember (-8.5, 113.5), Besuki (-7.5, 113.5) and Tuban (6.5, 112.0).

3 Result and Discussion

The four selected locations showed a noticeable different waves heights and periods. Waves on the north coast (Tuban and Besuki) had low significant wave height (less than 1 m) and small period (2-8 seconds). On other hand, in the south coast (Pacitan and Jember), waves had a higher significant wave height (up to 4 m) with a larger period (10 - 16 seconds). Wave spectrum for the four locations is shown in figure 2.

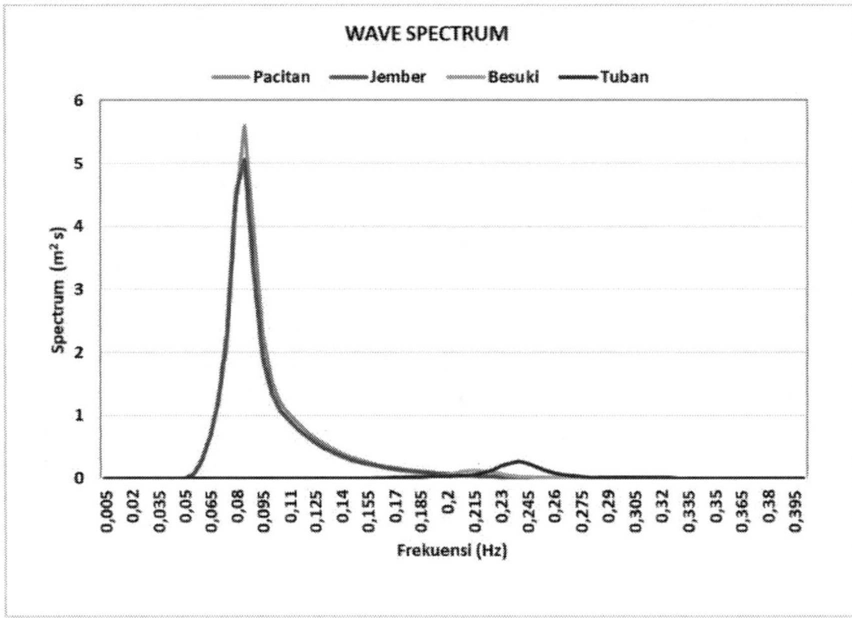

Fig. 2. Wave Spectrum.

Figure 3 – figure 6 show the waves energy flux which was calculated using equation (1). In Pacitan Coast, maximum wave energy was 232 MWh/m/year. Furthermore, in Jember coast, maximum wave energy was 190 MWh/m/year. On the other hand, Besuki and Tuban only produce energy respectively 8.6 MWh/m/year and 9.5 MWh/m/ year. Thus the southern coast of Java produced energy of about 200 GWh per km of beach within a year.

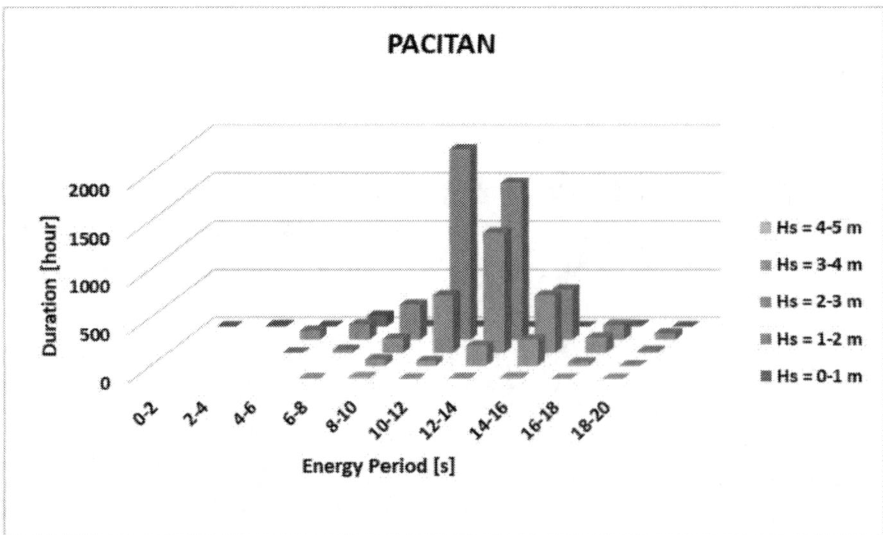

Fig. 3. Wave Energy in Pacitan

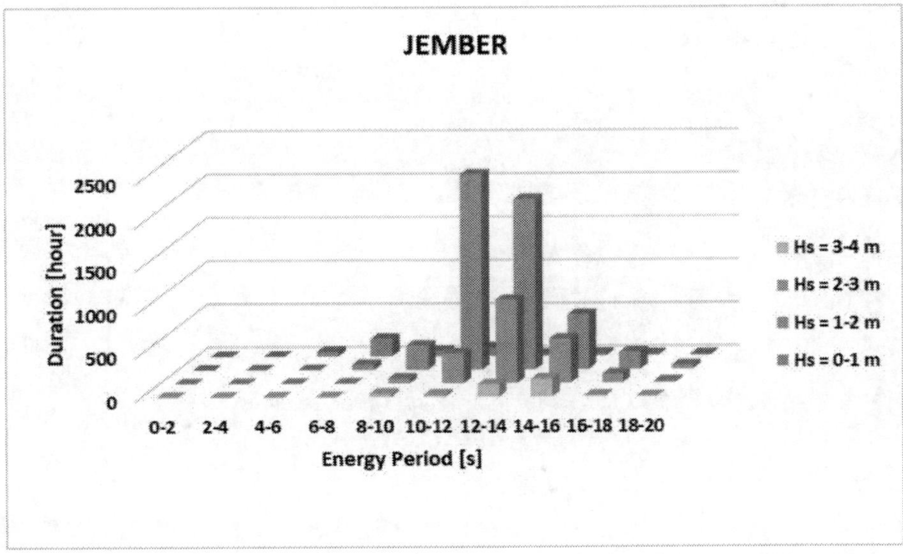

Fig. 4. Wave Energy in Jember

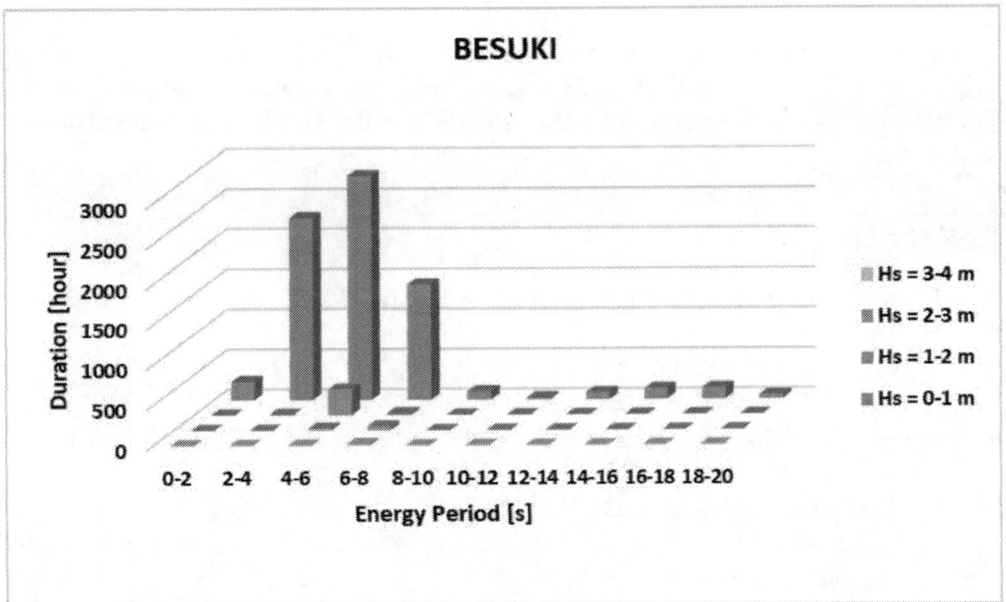

Fig. 5. Wave Energy in Besuki

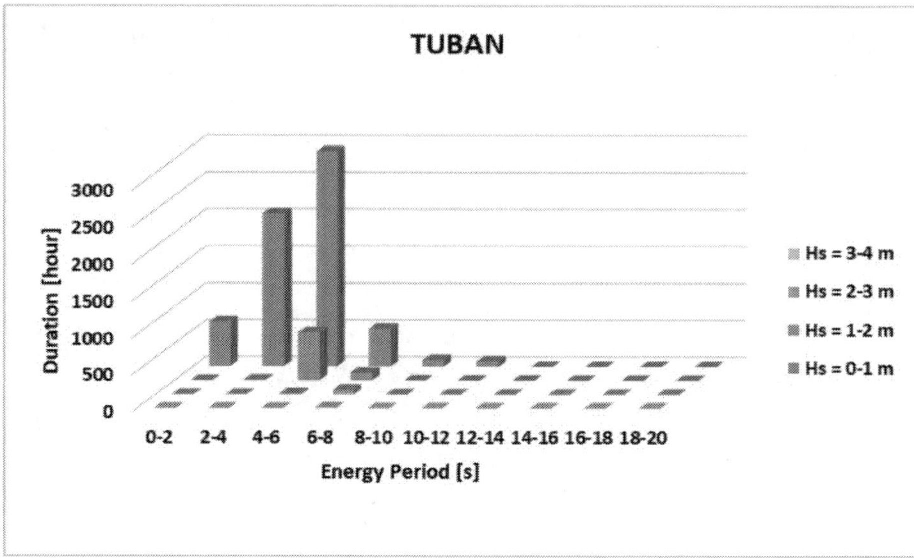

Fig. 6. Wave Energy in Tuban

4 Conclusion

The findings of the study could be revealed that the highest potency of wave energy to be converted into electrical energy in East Java occurred in the south coast of East Java. The coast of Pacitan had energy of 232 MWh/m/year and the coast of Jember had energy of 190 MWh/m/year.

Reference

1. D. Nugroho Sugianto, M. Helmi, I. Alifdini, L. Maslukah, S. Saputro, M. Yusuf, and H. Endrawati, IJMET, **8**, 448–459 (2017)
2. D. Anggraini, M. I. A. Hafiz, A. F. Derian, Y. Alfi, IJST, **1**, 228–239 (2015)
3. A. Cornett, Proc. ISOPE, 8, 318–326 (2008).
4. B. Drew, A. R. Plummer, and M. N. Sahinkaya, Proc. ImechE. Part A: J. Power and Energy, **223**, 887–902 (2009)
5. J. Falnes, Marine Structures, **20**, 185–201 (2007)
6. R. Sandro, Arnudin, A. Tussadiah, R. M. Utamy, N. Pridina, and L. N. Afifah, Energy Procedia, **47,** 242-249 (2014)
7. N. P. Purba, J. Akuatika, **5**, 8–15 (2014)
8. L. S. Blunden, A. S. Bahaj, and N. S. Aziz, Renew. Energy, **49**, 137–142 (2013)
9. M. Gonçalves, P. Martinho, and C. Guedes Soares, Renew. Energy, **68**, 774–784 (2014)
10. P. Mota and J. P. Pinto, Renew. Energy, **71**, 8–17 (2014)
11. R. Atan, J. Goggins, and S. Nash, Energies, **9**, 1–29 (2016)

Bio-Inhibitor on Corrosion Rate of ASTM A53 Steel in Marine Environment

Herman Pratikno[1*], *Harmin* Sulistiyaning Titah[2] , and *Yossy* Kurnia Shandy[1]

[1]Department of Ocean Engineering, Faculty of Marine Technology, Institut Teknologi Sepuluh Nopember, Keputih, Sukolilo, 60111 Surabaya, Indonesia
[2]Department of Environmental Engineering, Faculty of Civil, Environmental and Geo Engineering, Institut Teknologi Sepuluh Nopember, Keputih, Sukolilo, 60111 Surabaya, Indonesia

Abstract. There are many methods to prevent the corrosion process. One of them is corrosion inhibitors usage. Inhibitor divided into two types, there are organic inhibitor and non-organic inhibitor. Organic inhibitor could be obtained by extracting some materials which exist in nature. In this research, bio-inhibitors were made from rambutan (*Nephelium lappaceum*) peel, banana (*Musa acuminata Colla*) peel, mango (*Mangifera indica L.*) leaves and papaya (*Carica papaya*)leaves. The determination of corrosion rate used weight loss method. The material of carbon steel ASTM A53 was used in the research. The salinity of artificial seawater was 35 ‰. The results showed that corrosion rate on carbon steel ASTM A53 was 0.2172 mpy at mango leaves and the efficiency of those inhibitor was 44.5% . The corrosion rate carbon steel ASTM A53 was 0.2436 mpy for corrosion rate and 37.8% of inhibitor efficiency when it used papaya leaves as bioinhibitor. Meanwhile, it reached 0.2739 mpy and 0.2812 mpy of corrosion rate for banana peel and rambutan peel, respectively. The efficiency of inhibitor were 30% and 28.2% for banana peel and rambutan peel, respectively. The corrosion rate on carbon steel ASTM A53 without inhibitor was 0.3914 mpy. Based on the microstructure results, all specimens in all variation of treatments showed pitting corrosion and uniform corrosion on specimen. In conclusion, mango leaves, papaya leaves, banana peel and rambutan peel have potentially to be used as bio-inhibitor to decrease the corrosion rate in marine environment.

1 Introduction

Corrosion can cause disastrous damage to metal and alloy structures causing economic consequences in terms of repair, replacement, product losses, safety, and environmental pollution. Due to these harmful effects, corrosion is an undesirable phenomenon that ought to be prevented [1].There are several steps to inhibit the process of corrosion, one of them is the inhibitors. The inhibitor may be either organic or inorganic. It serves to add to the metal material in a corrosive environment so that it will inhibit and lower the rate of corrosion. In recent days, many alternative ecofriendly corrosion inhibitors have been

* Corresponding author: hermankelautan@gmail.com

developed, and they range from rare earth elements to organic compounds [2] Organic inhibitors can be obtained by extracting some of the ingredients found in nature. Some examples of natural materials that can be used as an organic inhibitor of banana peel, orange peel, dragon fruit peel, mangosteen peel, mango leaves, papaya leaves, and so forth. The concentration of nature inhibitor obtained results that vary depending on the type of material used. Inhibitors decrease or prevent the reaction of the metal with the media. They reduce the corrosion rate by (i) adsorption of ions/molecules onto metal surface, (ii) increasing or decreasing the anodic and/or cathodic reaction, (iii) decreasing the diffusion rate for reactants to the surface of the metal, (iv) decreasing the electrical resistance of the metal surface. (v) inhibitors that are often easy to apply and have in situ application advantage [3].

Some of the ingredients were in nature inhibitor include tannins, alkaloids, saponins, amino acid pigment, catechins, and proteins that have the ability to reduce the rate of corrosion. Tannins and their derivatives can be used to protect steel, iron, and other tools from corrosion [4]. Extracts of tobacco from twigs, stems, and leaves can protect steel and aluminium in saline solutions and strong pickling acids [1,5]. According to Paul and Koley [2], corrosion inhibition of carbon steel by few green inhibitors viz. garlic, yeast, pepper, and coffee or green inhibitors have been found to decrease the acid corrosion of mild steel by several folds with about 70–90% adsorption on the surface. The other study reported that watermelon extract has been found to have strong effect on corrosion of mild steel in HCl and H_2SO_4 [6]. The inhibition efficiency increases with the concentration of the extract increases.

Although, some research reported the usage of many nature inhibitor but it is rare for carbon steel in marine environment. The aim of the research were to determine of corrosion rate and inhibitor efficiency of ASTM A53 steel after add with natural inhibitor from banana peel extract, rambutan peel, mango leaves, and papaya leaves.

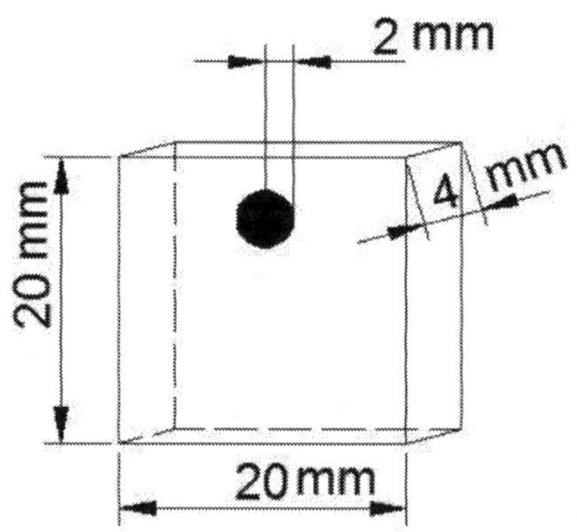

(a) **(b)**

Fig. 1. (a) Dimension of specimen and (b) the specimen was ready to be used

2 Materials and methods

2.1 Preparation of Specimen

A steel pipes of ASTM A53 specifications were used in this research [7] and the preparation of specimens was conducted based on the American Society of Mechanical Engineers (ASME). Figure 1 showed the demension of specimen (20 x 20 x 4 mm) and the appearance of specimen after it through some process of preparation based on ASME. The following steps were conducted:

a. Cut-shaped material with dimensions of 20 mm x 20 mm x 4 mm.
b. Drilling was carried out on the upper side portion of the specimen.
c. Finishing the surface of the specimen with grade # 80 sandpaper to # 2000.
d. The process of pickling the specimen was soaked using low grade HCl with ASTM A380 standard.
e. After that, the specimen was cleaned by rinsing with running water then dried and ready for use.

2.2 Preparation of Bio-Inhibitor

About 300 grams of every basic ingredients of natural inhibitor put on erlenmeyer. For extraction solvents using N-Hexan as much as 500 ml of each ingredient. The method for the manufacture of inhibitors was reflux or upright coolant. The preparation of the extract started by mixing the material with the solvent on the erlenmeyer tube. After that, it installed an upright cooling extraction tool and heated the mixture of the ingredients and solution over the hotplate for approximately 1 hour. The color of solution changed due to the mixing of the compound in the inhibitor with the solvent. After that, the solution was filtered with filter paper to separate the liquid and solids of the inhibitor material. The solution was cooled by kept it in room temperature. Then the resultant solution was heated to evaporate the N-Hexan solution to obtain a concentrated inhibitor extraction. After obtaining a thicker solution, the next step was cooled by immersion for some time. The inhibitor was ready to be used on the specimen.

2.3 Preparation of Artificial Seawater

In this research, the artificial seawater with salinity of 35 ‰ was used due to the salinity was a salinity in the depths of the intermediate ocean. The standard to preparation of artifisial seawater was ASTM D1141-98 [8].

2.4 Immersion Testing

The specimen was tested by immersion tehnique in a prepared seawater solution with salinity 35‰ using ASTM G31-72 standard [9]. There were two different treatments, namely treatment with bio-inhibitor addition and treatment without addition of bio-inhibitor to the specimens. Testing was carried out for 2 weeks. The specimens were be sprayed with each solution extraction of rambutan peel, banana peel, mango leaves and papaya leaves. One specimen was be sprayed with 10 mL of solution extraction. Immersion testing was conducted in plastic glass with size of 300 mL and the artificial seawater was 250 mL for each plastic glass.

2.5 Calculation of Corrosion Rate and Efficiency of Inhibitor

Running rate of corrosion is a rapid propagation of material quality decline against time. There is a formula for calculating the corrosion rate based on the ASTM G1-03 standard [10] as follows:

$$Corrosion\ Rate\ (mpy) = \frac{K\ x\ W}{A\ x\ T\ x\ D} \quad (1)$$

Where :
K = Constanta
T = Time of eksposure (h)
A = Surface area (cm^2)
W = Weight loss (gram)
D = Material density ($gram/cm^3$)

In addition, the efficiency of inhibitors is also one of the important aspects of this study. It is the value that determines how optimal an inhibitor works. Efficiency of inhibitors can be calculated using the following equation:

$$Efficiency\ of\ inhibitor(\%) = \frac{V1}{V0}\ x\ 100 \quad (2)$$

Where :
V1 = Corrosion rate with addition of inhibitor
V0 = Corrosion rate without inhibitor

2.6 Microstructure Testing

After corrosion testing was conducted, the microstructure of specimens were determinated using a microscope for detailed morphology of the specimen structure. It was used for documentary evidence and it can be known that the specimen differences between before and after testing.

3 Results and Discussion

3.1 Extract of Rambutan Peel, Banana Peel, Mango Leaves, and Papaya Leaves

Rambutan peel extraction had a physical feature of a yellowish solution. The smell of rambutan peel extraction was also not too strong. The application of bio-inhibitor to the specimen was conducted using spray method. The solution extraction of rambutan peel was sprayed to the surface of specimen. The bio-inhibitor color after it was sprayed to the specimen was not very visible.

Banana peel extraction had the physical characteristics of brownish yellow solution. However, when compared with rambutan peel extraction solution, the color of banana peel extraction had more concentrated. There was a distinctive aroma from banana peel extraction. However, there was a sap in banana peel extraction due to the specimen was sticky when it held.

Solution of mango leaves extraction had the physical characteristics of dark green solid solution. When compared with other extractions, the color of mango leaves extraction was concentrated. The aroma of mango leaves extraction was quite strong. The appearance of color was dark green when the extraction of mango leaves was sprayed to the specimen.

Solution of papaya leaves extraction had physical characteristics similar to mango leaves extraction. Solution of papaya leaves extraction was dark green. When compared with other extractions, the color of the papaya leaves extraction was concentrated. The aroma of the papaya leaves extraction was quite strong as the aroma of green tea. Figure 2 showed the appearance of solution extraction of mango leaves, banana peel, papaya leaves and rambutan peel.

Fig. 2. the appearance of solution extraction of mango leaves, banana peel, papaya leaves and rambutan peel.

3.2 Corrosion Rate

Figure 3 showed that mango leaves has the lowest corrosion rate on ASTM A53 in 35‰ of salinity after immersion test for 14 days. It reached 0.2172 mpy. The corrosion rate on ASTM A53 with addition of papaya leaves, banana peel and rambutan peel solution extraction were 0.2436, 0.2739, and 0.2812 mpy, respectively. The corrosion rate on ASTM A53 without nature inhibitor or bio-inhibitor addition was 0.3914 mpy. Based the results, it showed that solution extraction of mango leaves, papaya leaves, banana peel and rambutan peel potentially to be used as bio-inhibitor to descrease the corrosion rate on ASTM A53 material in salinity of 35‰. Mango leaves extract and papaya leaves can inhibit corrosion rate better than rambutan peel extract and banana peel extract. This due to solution extraction of mango leaves and papaya leaves have stronger adhesion on the surface of the specimen. Besides that, solution extraction of banana and rambutan peel were more easily react with corrosive media testing i.e. NaCl solution 35 ‰ so that those were more quickly decomposes.

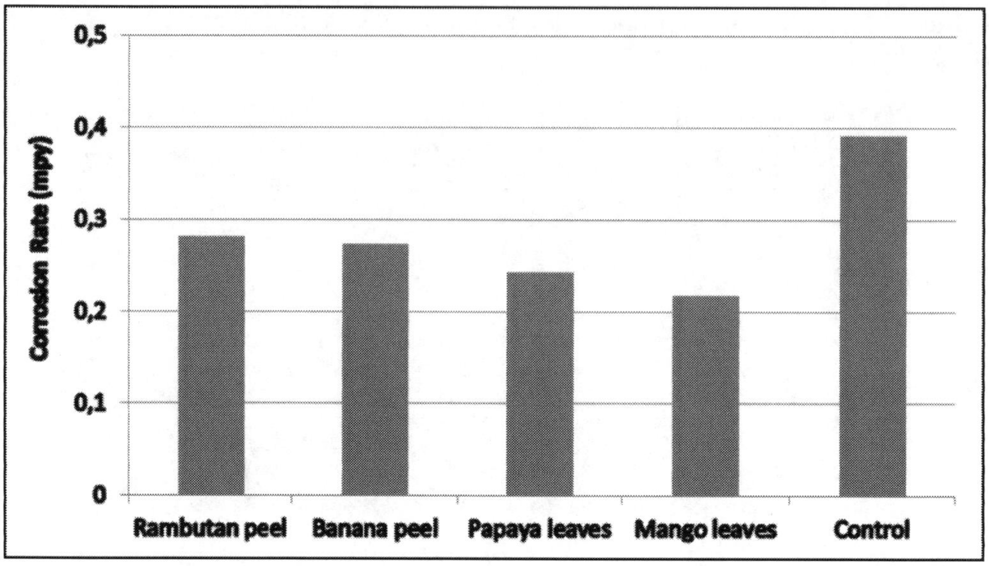

Fig. 3. Comparison of corrosion rate on ASTM A53 material.

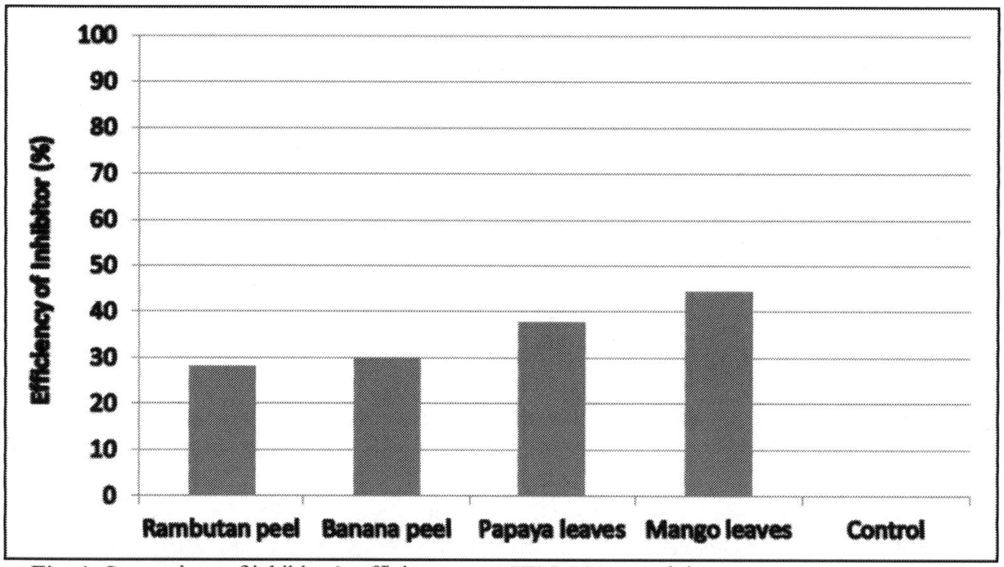

Fig. 4. Comparison of inhibitor's efficiency on ASTM A53 material

3.3 Efficiency of Bio-Inhibitor

The efficiency of mango leaves as inhibitor was 44.5% (Figure 4). Meanwhile, others nature inhibitor i.e papaya leaves, banana peel and rambutan peel can be used as inhibitor although the efficiency were below the mango leaves. The reduction of corrosion rate was be predicted the inhibitor or solution extraction coated the surface of the metal surface so that it could inhibit the occurrence of redox reactions in the corrosion process. According to Rani and Basu [3], one of mechanism of reduction corrosion rate was decreasing the anodic and/or cathodic reaction. The ability of mango leaves, papaya leaves, banana peel and rambutan peel were as inhibitor on corrosion due to those plants contained tannin and saponin. According to Ningsih et al. [11], extract of mango leaves contained tannin and saponin. However, papaya leaves, banana peel and rambutan peel also contained tannin and saponin, but it suggested the concentration below the mango leaves. It indicating that mango leaves has a high potentially to be used as a corrosion inhibitor on ASTM 53 in marine environment.

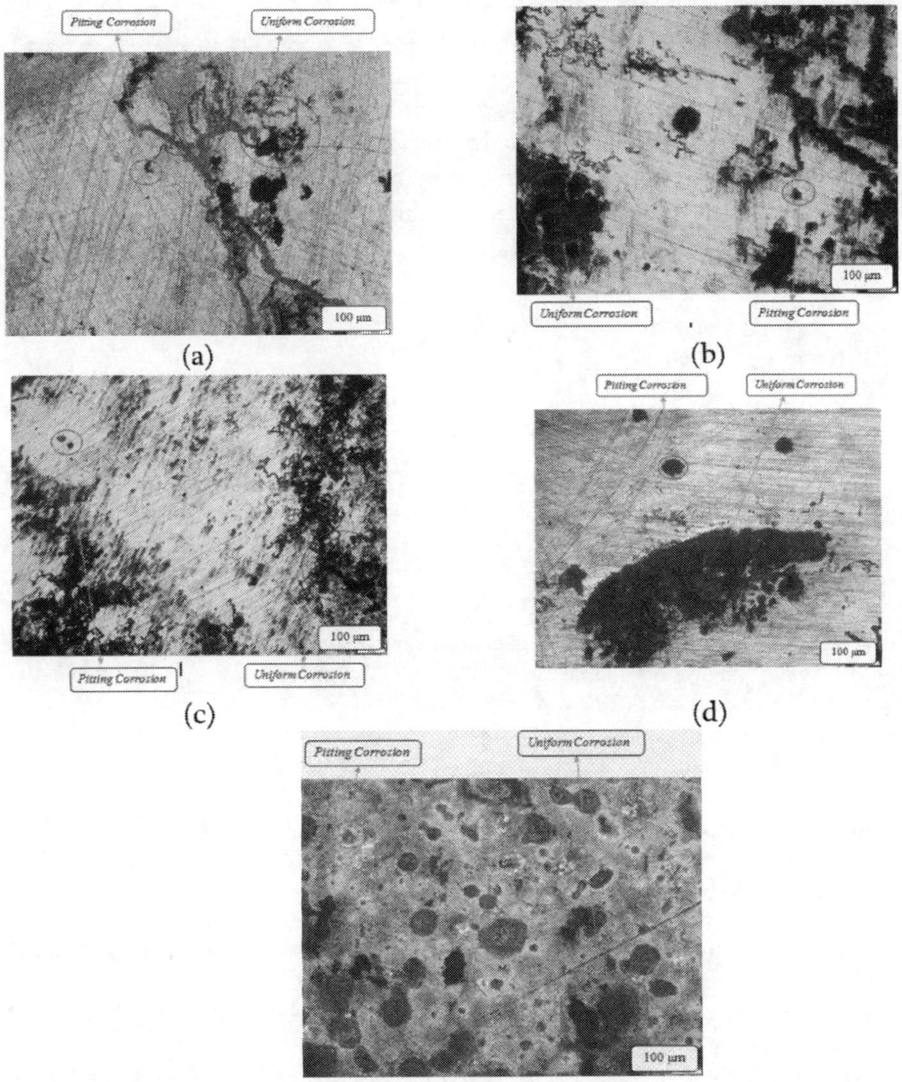

(e)

Fig. 5. Microstructure on ASTM A53 material after addition with microstructure (a) solution extraction of mango leaves (b) solution extraction of papaya leaves (c) solution extraction of banana peel (d) solution extraction of rambutan peel (e) control or without any addition of bio-inhibitor.

3.4 Microstructure

The microstructure in the following figure with 100x magnification consists of light-colored ferrite grains and dark pearlit grains. Grain ferrite was a structure in steel that tends to be subtle. The characteristics of pearlit grains tend to be more rough when compared to grains ferlit. Figure 5 showed the microstructure of specimens in control or without addition of solution extraction and in various of bio-inhibitors with 100x magnification.There was a difference in the microstructure of the intermediate specimen when it comparated between control and with addition of solution extraction. The corrosion on specimen could be detected through black holes or dark patches in photo of microstructure. The photo of microstructure can identify the type of corrosion that occurs in a specimen. The majority of corrosion types that were occured on the test specimens were pitting corrosion or corrosion hole. This pitting corrosion was be depicted a dark hole in the microstructure. In addition, uniform corrosion were be detected on some surface of specimen. However, the microstructure results showed only a part of specimen. Based on Figure 5, all specimens in all variation of treatments showed pitting corrosion and uniform corrosion on specimen.

4 Conclusion

Solution extraction of mango leaves, papaya leaves, banana peel and rambutan peel can be used as nature inhibitor on preventing of corrosion on ASTM A53 in salinity of 35‰. The efficiency of solution extraction of mango leaves, papaya leaves, banana peel and rambutan peel as bio-inhibitor on corrosion were 44.5; 37.8; 30 and 28.2%, respectively.

References

1. J. Buchweishaija. *Phytochemicals as green corrosion inhibitors in various corrosive media a review*. Chemistry Department, College of Natural and Applied Sciences, University of Dares Salaam. (2009)
2. S. Paul, I. Koley. Journal of Bio- and Tribo-Corrosion **2(6)** 1-9 (2016)
3. B.E.A. Rani, B.B.J Basu. Int J of Corrosion 1-15 (2012) doi:10.1155/2012/380217
4. N. Patni, Agarwal, P. Shah. Chinese J of Engineering 1-9 (2013)
5. G.D. Davis, J.A. Von Fraunhofer, L.A Krebs, C.M. Dacres. Corrosion 1558 (2001)
6. N.A. Odewunmi, S.A. Umoren, Z.M Gasem. J Ind Eng Chem. (2014) doi:10.1016/j.jiec.2014.02.030
7. American Society for Testing and Materials. ASTM A53 Standard *Spec. for Pipe, Steel, Black and Hot Dipped, Zinc Coated, Welded and Seamless*. ASTM Society (1996)
8. American Society for Testing and Materials. ASTM D1141-98 *Standard Practice for Artificial Sea Water*. ASTM Society (2004)
9. American Society for Testing and Materials. ASTM G31-72 *Standard Practice for Laboratory Immersion Corrosion Testing of Metals*. ASTM Society (2004)
10. American Society for Testing and Materials. ASTM G1-03 *Standard Practice for Preparing, Cleaning, and Evaluating Corrosion Test*. ASTM Society (2002)
11. D.R. Ningsih, D. Zusfahair, Mantari . Jurnal Kimia Riset **2(1)** 61-68 (2017)

The Estimation of Volume and Type of Household Waste in the Coastal Village, Besuki Region, Situbondo, East Java

Dina T Sseptiningtiyas[1,*], *Wazirotus* Sakinah[1] , and *Ani* Listriyana[1]

[1]Universitas Abdurachman Saleh, Situbondo, 68321, Indonesia

Abstract. Population growth is very influential on the amount of waste generated. The available land is very limited. The coastal community in Besuki mostly disposes garbage in the sea of any river due to unavailability of landfill. So we need a study on the estimated volume and type of waste generated by households in coastal village in Besuki region, Situbondo. This study used 10 household samples spread over coastal village. The study was conducted for 8 days using survey method with sampling (Proportional Random Sampling). Based on the results of the preparation, obtained the waste generated in the form of organic waste and non-organic waste. The composition of organic waste can be selected into several types of leaves, garbage, household, food scraps, and wood. As for the type of non-organic waste can be sorted into plastic, paper, metal, rubber, styrofoam, and cloth. The total amount volume of waste produced per household in a day in coastal villages is 47,3 liters with the largest composition of waste generated from plastic by 43,36%.

1 Introduction

The accumulation of discarded objects in the environment has become an issue of significant global concern. On land and in freshwater habitats, these objects are typically referred to as trash, and in the marine environment, these objects are referred to as marine debris. Trash and marine debris affect aesthetics and/or aquatic life across every habitat they touch from watersheds to the ocean [1].

Waste is a material that is wasted or removed from the results of human and natural activities that have not had economic value. Trash comes from households, farms, offices, companies, hospitals, markets. Household waste such as batteries, electric lights, electronics, pesticide packaging, clothing bleach, floor cleaners, paints, pressure tins (aerosols) are not only harmful but also toxic called Household Hazardous Solid Waste. The remaining drugs have potential health and environmental impacts. The greater the quantity of SB3-RT burned indiscriminately and disposed directly into the environment, the greater the risk of health problems and environmental pollution [2].

* Corresponding author: tiyasdina6@gmail.com

The amount of garbage that builds up is one of the serious problems for decades and can not be solved properly. Garbage generated increasingly day by day, it is influenced by the increasing of population. Domestic waste is a waste produced by people who have various activities. Data on waste generation is essential in selecting the type or type of equipment used in waste transport, the design of waste processing systems, waste treatment facilities, and landfill design. Determination of waste generation is usually expressed in volume and weight. The composition of waste is a representation of each component contained in the waste and its distribution. Usually expressed in weight percent (% by weight). This data is important for evaluating the necessary equipment, systems, programs and waste management plans of a city [3].

Waste problems can occur due to lack of land for waste shelters and lack of public awareness to dispose of waste in place. Some people throw garbage into rivers, seas, and some even burn them. It can damage the environment. Household waste can affect the quality of water. Resulting in pollution of water such as water that used for bathing and laundry water. Contaminated water can no longer be used for domestic use. Contaminated water and then can not be used again as a supporter of human life, will cause a very wide social impact and take a long time to recover, whereas the water is needed for household needs very much. [4]

One garbage problem occurred in the Pesisir village, Besuki region, Situbondo. Where in the area is still lack of polling stations or landfill, thus causing the community around to dispose garbage in the sea or river. The economic impacts of marine plastics on coastal communities are also considerable, especially for fisheries [5]. The purpose of this study is to determine the volume of waste per day generated by the coastal community, Besuki region, Situbondo.

Fig 1. Waste condition in Pesisir village

2 Study Area

This research was conducted in coastal area located at 70 43' South latitude and 1130 41' East longitude, Besuki region, Situbondo, East Java with 10,021 population people. Garbage condition in Pesisir village, Besuki region research sites.

3 Material and Method

The research method used is survey method with data sampling (Proportional Random Sampling) to compile estimation model of domestic waste volume. The sample used in this research is 10 samples of household waste spread in Pesisir village based on SNI-19-3964-1994 method. Here is the calculation done in this research: The number of sampling taken is done randomly with the amount as follows:

1. The number of instances of the soul and the head of household (KK) can be seen in Table 1 which is calculated based on the formula and 2 below [6].

   $$S = C_d\sqrt{P_s} \quad (1)$$

   Where:
   S = Number of instances (soul)
   C_d = residential coefficient
   C_d = Large city / metropolitan
   C = Medium / small / IKK
   P_s = Population (people)

   $$K = \frac{S}{N} \quad (2)$$

   Where:
   K = Number of instances (KK)
   N = Number of souls per family = 5

2. Estimation of domestic waste volume generated in each housing unit (ULP) [6]

   Calculate the average volume of domestic waste produced by each person on ULP-n

   $$v = \frac{\text{the total amount of domestic waste volume in all samples } ULP_{-n}}{\text{the average number of residents per mukim home on } ULP_{-n}} \quad (3)$$

   v = The average volume of domestic waste generated per person on ULP-n

   • Calculate the volume of domestic waste generated on ULP-n

   $$V = P \times v \quad (4)$$

   V = Volume of domestic waste generated on ULP-n

   P = Number of residents in ULP-n

   v = The average volume of domestic waste generated per person on ULP-n

Household waste sampling is done for 8 consecutive days. Trash then sorted according to the type of waste that is organic and non organic which then calculated the volume and type of waste generated equations should be centred and numbered with the number on the right-hand side.

4 Result and Discussion

Pesisir Village is one of the villages residing in Pesisir village, Besuiki region, Situbondo, East Java, Indonesia. The coastal village is 0.56 km^2 with a population of 10,021 inhabitants. The number of households in the Pesisir village are 3,932 [7].

Coastal is a village that is very close to the sea and very potential. It strategically easy to reach and find fish in the sea. Most of the population are working as a fisherman. Pesisir village is a village that located in the coastal area of Madura strait with the area divided into 6 (six) dusun, including dusun Petukangan, dusun Krajan, dusun Gudang, dusun Lesanan Kidul, dusun Lesanan Lor, and dusun Mandaran.

Before doing this research, we conducted a preliminary survey on the condition of waste in the coastal area. Trashes in the coastal area are generated from household waste, resident garbages whose open a shop, or factory waste such as cracker factory. Garbage in the coastal area is dumped in the sea or rivers due to the absence of temporary garbage dumps. In addition there is also waste that is burned by the surrounding community.

Fig 2. Waste condition in Pesisir village

There are number of points which solid waste may be come in relation with living organism such as soil adsorption, storage and biodegrading, plant uptake, ventilation, leaching, insects, birds, rats, flies and other animals. Direct dumping of untreated waste in seas, rivers and lakes results in the plants and animals that feed on it [8]. Waste discharged into the water can damage the condition of water or marine biota. Most of the waste that are disposed in the form of plastic waste that can not decompose in a short time. Organic waste that is wasted into marine waters when exceeding the ability of sea assimilation, can pollute the waters and cause excessive fertilization (eutrophication). This phenomenon will cause the decrease in dissolved oxygen levels due to the explosion of a population of certain organisms that can cause death of some other organisms in the waters [9].

As for inorganic waste that is difficult to decipher, marine organisms have the ability to receive bioaccumulation of metals in their bodies. But in the long time, this bioaccumulation will affect the oragnism was called subletal effects. In addition, if the organism is consumed by humans, it can cause serious poisoning [9].

Based on the survey results, most of communities around the coastal don't know about the impact of throwing the garbage into the waters espesially in the sea. So that's what makes coastal people throw garbage into the sea or river. In addition, economic factors are the reason why they don't use the transportation to bring the waste from the Public Works Office prepaid. In this study used 10 samples of household waste were taken for 8 days. The garbage collection was done on October 2, 2017 until October 9, 2017. The sample selection was spread in every hamlet of Pesisir village with sampling in each hamlet was done randomly. In the process of taking garbage is done every morning. Then the volume of waste and weight of waste generated are calculated.

We separated the garbage that was collected before calcalute the garbage volume. It was separated between organic waste and non organic waste. The kind of organic waste are divided into leaves, household waste, food scraps, and wood. While non-organic waste like plastic, paper, metal, rubber, styrofoam, and cloth. Then we calculated the garbage volume after it had been sorted.

Fig 3. Household waste that has been collected

From the results of research that conducted based on data collection methods sampling in the village of coastal Situbondo, we obtained data on the volume amount of organic and non organic waste.

Table 1. Percentage of organic garbage volume

Organic Garbage Type	Day (Waste volume and liters unit)								Total (l)	Percentage
	1	2	3	4	5	6	7	8		
Leaves	4,14	1,87	4,15	5,53	5,62	1,09	0,74	2,70	25,84	6,83%
Household waste	3,12	4,85	3,46	5,19	6,92	3,77	1,77	3,46	32,54	8,60%
Food scraps	2,42	3,46	2,08	2,77	4,42	3,12	3,81	1,04	23,12	6,11%
Wood	1,04	0	0,69	1,04	0,69	0	0,8	1,38	5,64	1,49%
Totals	10,72	10,18	10,38	14,53	17,65	7,98	7,12	8,58	87,14	23,03%

Based on the table above, it can be seen the volume of organic waste that produced for 8 days, is 25,84 liters of leaves, 32,54 liters of household waste, 23,12 liters of food waste, and 5,64 liters of wood waste. The total amount of organic waste produced by 10 samples was 87,14 liters. The volume of organic waste per day per household in Pesisir village is about 10,89 liters, so that the volume of organic waste produced by the Pesisir village for a day is 42.819,48 liters.

Table 2. Percentage of non organic garbage volume

Non Organic Garbage Type	Day (Waste volume and liters unit)								Total (l)	Percentage
	1	2	3	4	5	6	7	8		
Plastic	32,89	25,77	23,67	25,6	20,85	11,77	10,96	12,54	164,05	43,36%
Paper	10	11,89	15,38	9,68	10,27	7,52	6,61	8,12	79,47	21,00%
Rubber	0,69	3,46	2,08	1,73	2,42	0,69	1,77	1,38	14,22	3,76%
Sterofom	1,77	0,69	2,42	0,69	1,77	1,38	0,8	1,49	11,01	2,91%
Metal	3,81	1,77	1,73	2,42	0	0,69	1,04	0	11,46	3,03%
Cloth	2,77	0,69	0	1,77	1,38	0	2,19	2,22	11,02	2,91%
Total	51,93	44,27	45,28	41,89	36,69	22,05	23,37	25,75	291,23	76,97%

Based on the table above, it can be seen the volume of non-organic waste generated for 8 days of plastic waste of 164,05 liters, 79,47 liters of paper waste, 14,22 liters of rubber, 11,01 liters styrofoam trash, 11,46 liters of metals and fabric waste of 11,02 liters. The total amount of non organic waste produced by 10 samples is 291,23 liters. The volume of non-organic waste per day per household in besuki village is about 29,12 liters, so it can be obtained the volume of non-organic waste produced by the Besuki village for a day is 114,499.84 liters. Non-organic waste is the largest waste produced by the coastal villages.

Table 3. Estimated amount of garbage per household per day in Pesisir village

Category	Total of garbage (l)
Per household	47,3
Per coastal village	185.983,6

Based on the table above, the amount of waste per household in a day is 47.3 liters. With the number of households in the Coastal village of 3,932, there are about 185,983.6 liters of waste generated by the Pesisir village in a day. Plastic waste is the largest type of waste generated by coastal communities by 43.36% as illustrated in the following diagram. This is due to the type of plastic waste widely used in everyday life such as food and beverage wrap.

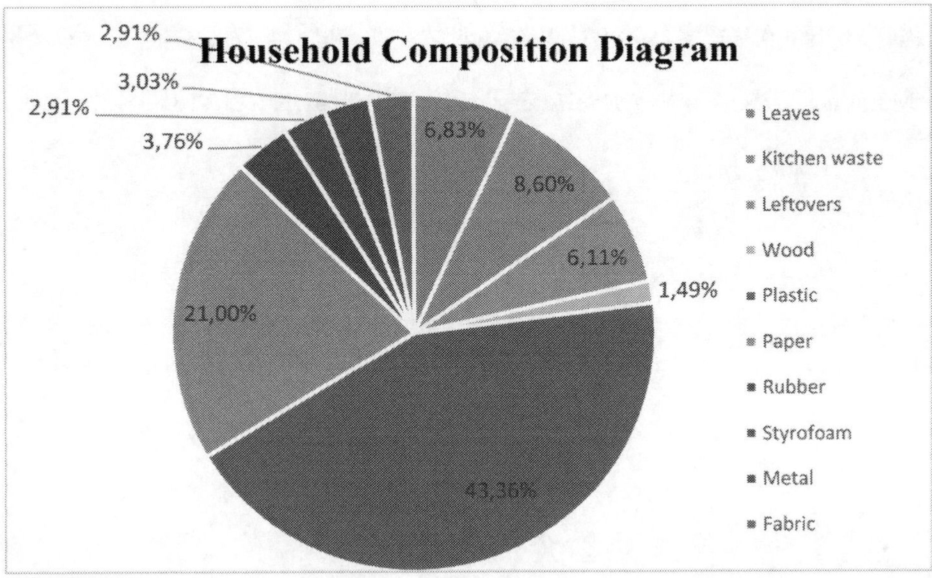

Fig 4. Household Composition Diagram

5 Summary

The waste in Besuki region was dominated by non organic waste with the waste mass in each household is 1.047,32 g/day. The estimation waste mass in Besuki coastal region is 167,57 g/people.day. Waste management in Besuki Coastal Region has percentage 73% citizen throw their wastes in the sea or/and the river, 27% has been burned, and 0% in landfill. Generally, they does not have the basic knowledge about marine debris impact with percentage up to 80% and the citizen who know about the knowledge is 20%. Their interest about self waste management is only 33%.

References

1. S.y Moore, M. Sutula, T. V. Bitner, G. Lattin, K. Schiff, *Trash and Marine Debris* (SCCWRP Technical Report 928, California, 2016)
2. Iswanto, Sudarmadji, E. T. Wahyuni, A.H. Sutomo, HMWMC, **23** (2016)
3. E. Damanhuri, *Waste Minimization as Solution of Municipal Solid Waste Problem in Indonesia* (The 6th ASIAN Symposium on Academic Activities for Waste Management, Padang- Indonesia, 2004)
4. R, Hasibuan, AHW, **64** (2016)

5. F. Thevenon, C. Carroll and J. Sousa, *The Characterization of Marine Plastics and their Environmental Impacts, Situation Analysis Report* (International Union for Conservation of Nature, Switzerland, 2014)
6. SNI-19-3964-1994, *Metode pengambilan dan pengukuran contoh timbulan dan komposisi sampah perkotaan* (Badan Standardisasi Nasional, 2009)
7. BPS, Kabupaten Situbondo, *Kecamatan Besuki Dalam Angka 2017* (BPS Situbondo, Situbondo, 2017)
8. P. A, Kafeel A, *Impact Of Solid Wate On Health And The Environment*, **2315-4721** (2013)
9. Mukhtasor, *Pencemaran Pesisir dan Laut* (PT PRADNYA PARAMITA, Jakarta, 2007)

Trial Test of Fish Attractor "Piknet" Device in Saltwater Fish Tank

Nurul Rosana[1,*], *Suryadhi*[2], *Safriudin* Rifandi[2]

[1]Department of Fisheries, Faculty of Engineering and Marine Science, Hang Tuah University, Indonesia
[2]Department of Electrical Engineering, Faculty of Engineering and Marine Science, Hang Tuah University, Indonesia

Abstract. The fish attractor called "Piknet" is an instrument designed based on sound wave to maximize the fishing activity by gill-netters. The sound wave are used in "Piknet" ranged from 100-1000 Hz. The "Piknet" test was conducted to know the range of sound wave which has well responses by salt water fish. The result showed that the sound wave ranged 500 – 1000 Hz is the best one.

1 Introduction

As Bregman points out, the world abounds with sound and animals use sound to glean a great deal of information about their environment [1]. Especially in sea, fish use sound to do so many activites supported their live : prelude or during mating, in warning of danger, and maybe in many other context that we have yet to understand. This fact, let human to improve a lot of technologies which can increase the fishing productivity in sea.

Fish's response to sounds are in very wide range of infrasonic, sonic, or ultrasonic [2]. One of the fishing technology developed by human is fish attractor. In Indonesia, this technology is starting to be studied and developed by both government and private parties. Some types of proven are pikat, alpine and electrofish [3]. In international issue, people are concern to the effects of human-generated sound on fish and other aquatic organisms by the influence of human-generated voices on fish and other aquatic organisms [4].

The fish attractor called "Piknet" is an instrument designed by wave sound based. This technology is intended to help gill-netters to increase their number of fish catching. Traditional fisherman in Indonesia are less of technological minded in their daily fishing. This lead in decreasing their income as less of fish that they can obtain daily. In the other hand, modern fisherman which have big of capitals do a huge fishing in many Indonesian areas [5].

Piknet can be one of solution to the problem and hopefully this device can be more developed in the near future. The sound wave used in "Piknet" ranged from 100 – 1000 Hz [3]. This paper will describe the experimental which was conducted to find the best responsive range of Piknet frequency for salt water fish.

* Corresponding author: nurul.rosana@hangtuah.ac.id

2 Research Method

Two range of frequency will be examined to 18 fish in salt water. The frequency ranged between 100-500 Hz and 500-1000 Hz. The selection of these frequency range is based on related literatures. The test was conducted 3 times in 15 minutes for each frequency range to observe the fish responses. Piknet device is immersed and installed on the right side of saltwater tank (Fig. 3).

Figure 1 show the device called Piknet. The cross-sectional dimension is 9 cm x 6 cm with 4.5 watts and 9 volt battery in power consumption. Figure 2 shows the step by step procedure of experiment.

Fig 1. The fish caller device "Piknet"

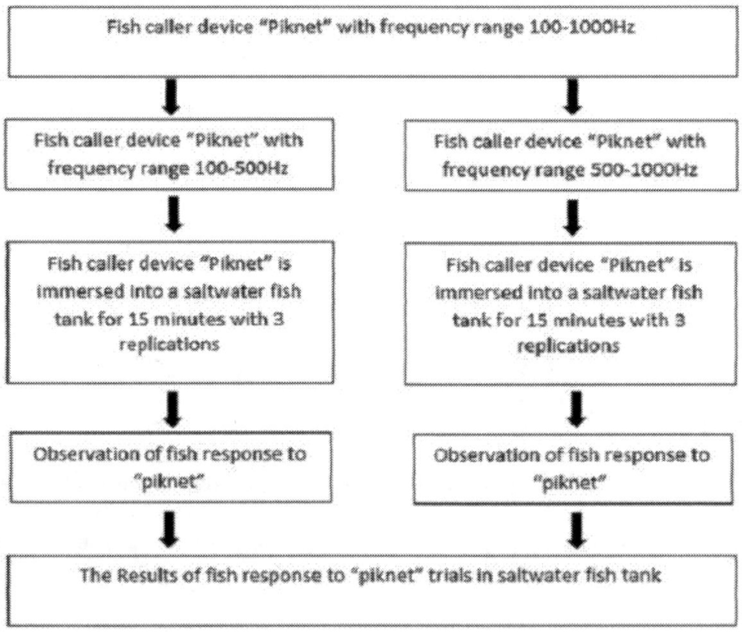

Fig 2. The experimental diagram

3 Result and Discussion

Experiment was conducted as the diagram showed in Figure 2. For frequency ranged 100 – 500 Hz, the fish do not have a significant response to come to Piknet after 10 minutes. Figure 3 shows that phenomena, that all of fish are in the left side or opposite of Piknet.

Otherwise, Piknet which have frequency range between 500 – 1000 Hz are very attractive for various types of fish after 2 minutes experiment (See Figure 4).

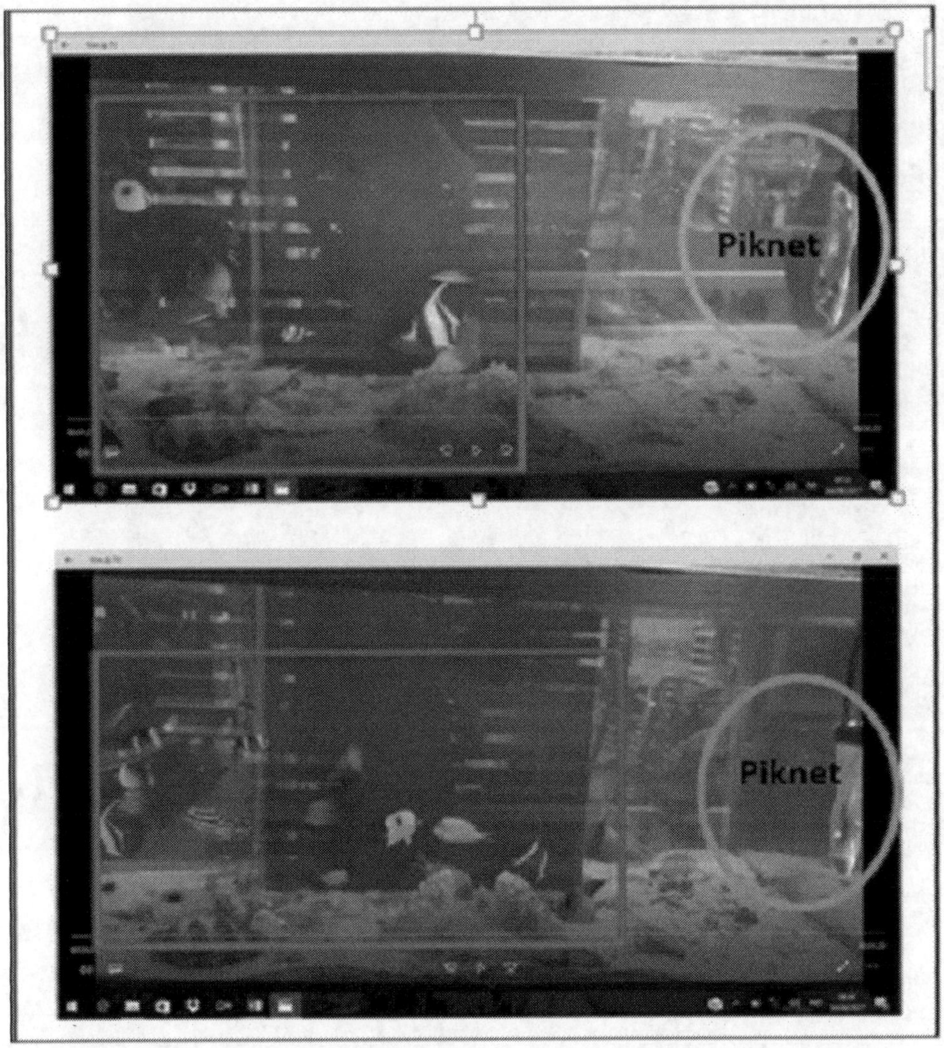

Fig 3. The movement of fish after 10 minutes "Piknet" at the frequency of 100-500Hz

Informations was brought by sound signals explains the state of threatening dangers, aggressive circumstances to frighten the enemy, or the call of favor [6]. Sound is also produced from other behavioral impacts such as eating, moving, avoiding enemies, and reproduction (sexuality and enlargement phases) [7]. From Table 1, it can be seen that the most responsive frequency is 500-1000 Hz. This frequency can attract 88% of fish in salt water tank, other wise only 22% of fish.

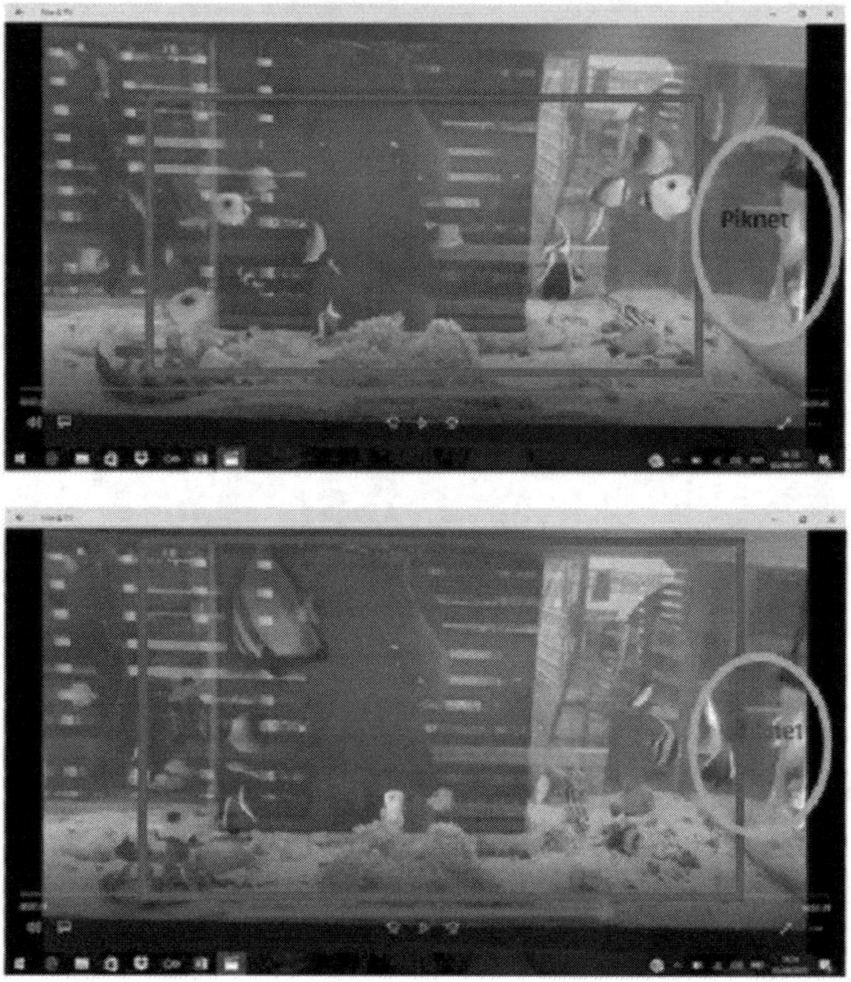

Fig 4. The movement of fish after 2 minutes "Piknet" at the frequency of 500-1000Hz

Tabel 1. Quantity and response of fish to "piknet"

Range of Frequency (Hz)	Quantity of Fishes in Saltwater Fish Tank	Quantaty of Fish Response	Percentage (%)	Fish response
100-500	18	4	22	Most of the amount of fish in the left side of the tank, opposite of Piknet.
500-1000	18	16	88	Most of the amount of fish begin to come to the "Piknet" that installed on the right of the tank, and rotate back to the other side then return closer "Piknet"

4 Conclusion

The experiment show that "Piknet" can work very well in the frequency ranged between 500 – 1000 Hz. This result is one of the step stone for further development toward establishing national fish attractor called "Piknet".

References

1. Bregman AS, *Auditory Scene Analysis: the Per-ceptual Organization of Sound*, MIT Press, Cambridge, MA, (1990)
2. G.V. Nikolsky, *The Ecology Of Fishes. Translet From Russian. L Birket. Academic Press. Inc. London*, (1963)
3. N. Rosana and Suryadhi, *Determination of Sound Waves for Design of "Piknet". Proceding of Hang Tuah University National Marine Seminars, Surabaya,* (2017)
4. A.N. Popper and M.C. Hastings, The Effects of Human-Generated Sound on Fish, Integrative Zoology, **4**, 1, pp. 43–52 (2009)
5. Hikmah, Mutiara, Jurnal H Int., **11**, (2013)
6. Pratt, Mary M, *Better Angling With Simple Science: The White Friars Press. London,* (1975)
7. A.N. Popper and C.Plat, *Inner Ear and Lateral Line P. 116 – 117. In David H. Evans. (ed). The Physiology Of Fishes. CRC Press. Boca Raton*, (1993)

Determination of shoreline change along the East-Java coast, using Digital Shoreline Analysis System

Eva Louise den Boer [1,*], Suntoyo [2], and *A C* Oele [3]

[1] Student, Delta Academy, HZ University of Applied Sciences, Vlissingen, the Netherlands. Research Intern, Ocean Engineering Department, Faculty of Marine Technology, Institut Teknologi Sepuluh Nopember (ITS), Surabaya, Indonesia (60111).
[2] Associate Professor, Ocean Engineering Department, Faculty of Marine Technology, Institut Teknologi Sepuluh Nopember (ITS), Surabaya, Indonesia (60111).
[3] Associate Professor, Delta Academy, HZ University of Applied Sciences, Vlissingen, the Netherlands.

Abstract. This study focused on determining the shoreline change at the East- Java coast by comparing the shorelines between 2006 and 2015, using digital shoreline analysis system (DSAS) techniques to identify and measure erosion (shoreline retreat) and accretion (shoreline advance). Shoreline change rates are calculated according to Shoreline Change Envelope (SCE), Net Shoreline Movement (NSM), End Point Rate (EPR), and Linear Regression Rate (LRR). Results present local variations in shoreline change distance as well as in shoreline change rates. The statistical methods showed similarities with the highest accretion distance of 991.57 meters in the region of Mulyorejo and highest erosion distance of 928.75 meters in Sidoarjo. The EPR shows a rate with a maximum of 169.98 m/year and a minimum of 3.428 m/year. As for the LRR a maximum of 189.58 m/year and a minimum of 5.467 m/year were conducted from calculations.

1. Introduction

A key requirement for effective coastal zone management is the analysis of changes in the past. Although changes along the coast over the coming century cannot be predicted simply by assuming the past trend will continue, it can contribute by constructing the basic knowledge used in future prediction formulas or models [1]. With a total length in coastline of 81 000 km spread over 17 508 islands, research in coastal development is valuable for disaster risk reduction and coastal management in the archipelagic country of Indonesia [2].

Located along the East Java Coast, the study area covers the areas of Pamurbaya and Sidoarjo. Geographically the districts are situated between 7°24' South Latitude - 112° 80' East Longitude and 7°48' South Latitude - 112°83' East Longitude. The region is known

* Corresponding author: Eva.denboer@hotmail.com

for its mangrove forests, marking the distinction between land and sea and forming a buffer zone between sea and shore. Along the coastal area of East-Java the mangrove forests form a unique ecosystem for marine species. The breeding and nursery grounds provided by the sheltered mangrove habitat directly relate to a productive source of economic benefits in sectors such as traditional fishery and aquaculture [3].
A common used program in shoreline change studies is the ArcGIS extension Digital Shoreline Analysis System (DSAS). The software program is used in both past shoreline changes [4] and in future prediction studies [5]. The main objective of this research is to determine the change in shoreline positions between 2006 and 2015 using DSAS as well as to identify erosion and accretion.

2. Theoretical background

2.1 Coastal Zone

The coastal zone is defined as the interface between the lithosphere, hydrosphere and atmosphere. It is a highly dynamic environment where physical processes play an important role in the development of the coast [6]. Geomorphological processes influencing the shaping of the coastal zone include the following:

- Waves are fluctuations on a water surface due to wind action. Within the coastal zone deep water regions change into shallow shore profiles where wave characteristics transform in height, speed and direction. The influence on coastal forms, whether changes will result in a movement seaward or landward, depends on the angle of the waves in comparison to the shore [7].
- Currents are the large masses of water moving in a specific direction. The movement is the result of differences in density, temperature and salinity of water between different locations [8]. The effect currents have on coastal morphology is little, except where warmer or colder water is brought; this changes ecological conditions which then influences shore dynamics [1].
- The tide is defined as the periodic rises and falls of large water bodies. Tidal influence is caused by the gravitational attraction between the Earth and the Moon. Contributing to the dispersion of sediments; the rising tides spread the sediment near the shore, whereas the falling of the tides contribute to erosion when sediments are taken away [1].
- Anthropogenic processes are caused due to human activities and occur in addition to natural forces. Rural populations have traditionally used mangrove trees to produce charcoal, fuelwood, boats and fish-traps. The main causes of loss of mangrove area have been overexploitation in combination with the development of shrimp farms. This results in serious human pressure on the ecosystem. When human activities increase the capability of mangrove forests to act as a natural buffer zone decreases, exposing the coastal zone to a greater vulnerability [9].
- Abrasion and accretion are the result of nearshore processes including beach morphology, wave climate, geology and human activities. The definition of accretion as stated by Thomas and Goudie [10] is the gradual increase in the area of land as a result of sedimentation. Erosion is defined as a group of processes whereby debris or

rock material is loosened or dissolved [10]. A dynamic equilibrium is reached when accretion and erosion are balanced, resulting in minimal changes [11].

- Sea level rise: in addition to manmade pressures, the Intergovernmental Panel on Climate Change (IPCC) forecasts a global rise in sea level, effecting mangrove forests worldwide [12]. The supply of sediments will determine the ability of mangroves to keep up with the rise in sea level, which indicates the vulnerability of that area [13].

2.2 Shoreline change

The boundary between land and sea, defined as the shoreline, changes variably due to multiple factors, which may be morphological, climatological or geological in nature. The geometry depends on the interactions between and among winds, waves, currents, and tides. The interactions vary locally, resulting in differences in erosion and accretion rates. The sedimentation process taking place causes the shoreline to move seaward. On the contrary, the process of erosion causes the shoreline to move landwards [14].

2.3 Mangrove area

Mangrove trees are defined as woody vegetation capable of growing in the intertidal zone along tropical and subtropical coasts. Their root structures have the ability of slowing down the process of erosion process by holding on the sediment particles. In addition, allowing the process of accretion to continue [15]. Because of this ability mangroves are used in afforestation programs where trees are planted in vulnerable, eroded areas to protect the coastal zone with a natural process [16].

2.4 Digital Shoreline Analysis System (DSAS)

Shoreline analyzation is done by using Digital Shoreline Analysis System (DSAS). This Arc Map extension, developed by US Geological Survey (USGS) is publicly available and can be downloaded at http://woodshole.er.usgs.gov/project-pages/dsas/.
Provided by the program, four methods were used as data analysis, including the SCE, NSM, EPR and LRR [17, 18].

3. Methodology

3.1 Literature study

To understand and analyse the problems stated in this research a literature study was performed. Background knowledge was gained from various scientific material, including journals, articles, books and internet websites.

3.2 Data collection

Data sets used in this research are as follows:

1. LANDSAT 7 satellite image data of the East-Java region with a resolution of 30 meters x 30 meters dated from 2006 to 2011.
2. LANDSAT 8 satellite image data of the East-Java region with a resolution of 30 meters x 30 meters dated from 2012 until 2015.

3.3 Data analysis

In this study, Digital Shoreline Analysis System (DSAS) is used to calculate change rates. A shoreline change estimation is obtained using two different kinds of statistical approaches: distance and rate. The Shoreline Change Envelope (SCE) and Net Shoreline Movement (NSM) represent the distance while the End Point Rate (EPR) and Linear Regression Rate represent the rate of shoreline change. SCE is the distance between the shoreline furthest from and closest to the baseline at each transect. It represents the total change in shoreline movement for all shoreline positions. NSM reports the distance between the earliest and most recent shoreline for each transect. It represents the total distance. EPR is calculated by determining the distance between the oldest and most recent shoreline in the data and dividing it by the number of years between them. This method provides the net rate of change over the long term. LRR is calculated using least squares regression lines from all shoreline positions along transect. It is generally used to express shoreline movement and the estimation of change rate.

4. Results and discussion

4.1 Results in shoreline change

In this research, the shorelines were extracted from Landsat multi-temporal satellite images every year from 2006 until 2015. The imagery was digitized and analysed using software programs ArcGIS and DSAS. To determine the shoreline changes a baseline was established with a buffer of 200 meters both seaward and landward parallel to the general shoreline orientation. As a second step, 1429 transect lines were set with a length of 12 km and 25 meter of space in between transect lines. With these settings the statistical methods SCE, NSM, EPR and LRR were calculated. The results of the statistical methods, calculated by DSAS are summarized in Table 1.

Table 1. Data collection of DSAS statistical methods.

Subject	SCE (m)	NSM (m)	EPR (m/y)	LRR (m/y)
Mean	165.53	30.89	3.43	5.47
Minimum	0	-928.75	-99.3	-129.53
Maximum	991.57	991.57	169.98	189.58

The detailed results are visualised in ArcGIS maps for the four statistical methods SCE, NSM, EPR and LRR. Five colour variations present the classified differences in erosion and accretion rates. The minimum value is shown in red and displays the most severe erosion as the green colour displays the maximum value: the highest rate of accretion. In between classes represent moderate erosion, neutral processes and moderate accretion with colours orange, yellow and light green. To prevent misinterpretation the choice of colour is chosen equally for all four statistical methods as viewed in Figures 1-4.

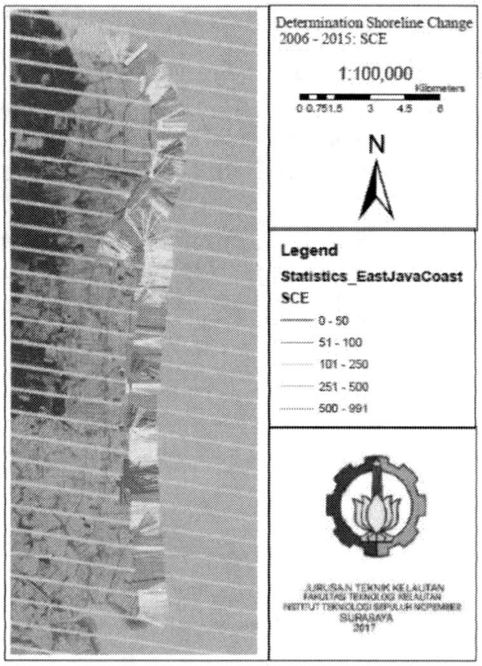

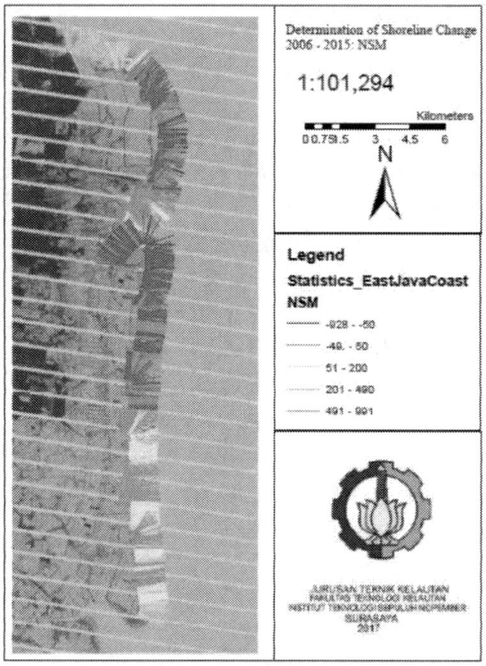

Fig. 1.(left) and **Fig. 2**. (right) Results of the Statistical methods SCE and NSM presented in an ArcGIS map.

The SCE value in Figure 1 represents the distance between the shoreline furthest from and closest to the baseline. Due to the extent of the study area and variations along the coastal areas, the shorelines defined as furthest from and closest to the baseline vary for each transect. The map shows a maximum distance of 991.57 meter, defining this as a maximum of accretion. The total distance in erosion and accretion in the period 2006-2015 is represented by the NSM value shown in Figure 2 The maximum erosion is -928.75 meter located in the Sedati district. The maximum accretion of 991.57 meter is located in the area of Mulyorejo.

The EPR analysis represents the speed of accretion and erosion in meter per year. Figure 3 presents the maximum accretion rate reaching 169.98 meter / year, while erosion rates reach a maximum of - 99.3 meter / year.

Figure 4 shows the LRR value expressing shoreline movement along the transect lines. With the Linear Regression Rate method, all shoreline positions are taking into account showing a maximum erosion rate of -129.53 meters between 2006 and 2015 as well as a maximum accretion rate of 189.58 meters.

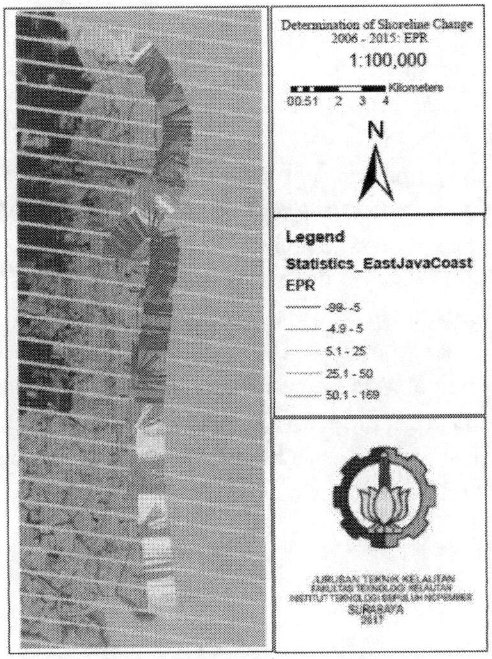

Fig. 3. Results of the Statistical method EPR presented in an ArcGIS map.

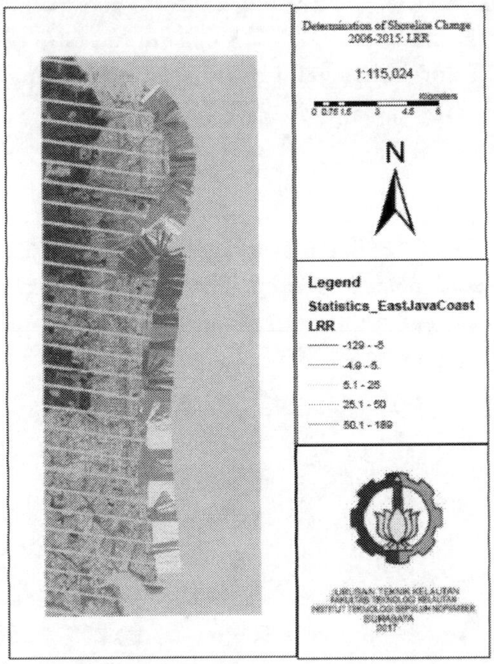

Fig. 4. Results of the Statistical method LRR presented in an ArcGIS map.

4.2 Discussion

In this study, the value of uncertainty needs to be considered when calculating shoreline change using DSAS. With the uncertainty value, natural and anthropogenic forces are taken into account (wind, wave, tidal, current, human influence) as well as the accuracy of measurement (digitation, interpretation, global positioning system error). Establishing a value of uncertainty is done by selecting a Confidential Interval of 95% as well as a shoreline uncertainty value of 5 meters. Special attention needs to be given to the tidal influence on the Landsat satellite images. It is taken into consideration that the shorelines may differ at different moments in time, due to the change in water level. Even though, because of the large extend of the study area, the tidal difference would only be visible to an inconsiderable amount. The possible fluctuations caused by the change in water level is taken into account in the uncertainty factor. Furthermore, the misinterpretation of the shoreline part of the uncertainty factor. The edge pixels are, when zoomed in, of insufficient quality for shoreline extraction increasing the chance of mistakes in characterization.

5. Conclusion

Results based on analysation and discussion show the differences in erosion and accretion rates. The ArcGIS maps for the SCE, NSM, EPR and LRR present local differences varying from severe erosion, moderate erosion, neutral, moderate accretion till severe accretion. Statistical methods EPR and LRR representing the rate show a continuous process of erosion or accretion over the study period. With continuing rates the maximum erosion and accretion distance (SCE and NSM) could likely increase in a linear way in the future.

Acknowledgements

The author shows her gratitude to Institut Teknologi Sepuluh Nopember (ITS) Surabaya, Indonesia and HZ University of Applied Sciences Vlissingen, the Netherlands, including lecturers, researchers and students for the contribution to the research and development in education. Special thanks is given to internship in-company supervisor Dr. Suntoyo, for giving the opportunity to present at ISOCEEN seminar.

References

1. E. C. Bird, *Coastal Geomorphology: an Introduction,* (2011)
2. S. Sukardjo, Integrated Coastal Zone Management (ICZM) in Indonesia, *Japanese of Southeast Asian Studies,* **40.2**, 200-218, (2002)
3. P. K. Asokan, Mangroves and its importance to Fisheries, (2012)
4. M. I. Joesidawati, Suntoyo, Shoreline change in Tuban district, East Java using geospatial and Digital Shoreline Analysis System (DSAS) techniques, *International Journal of Oceans and Oceanography,* 235-246, (2016)
5. N. K. Barman, S. Chatterjee, A. Khan, Trends of Shoreline Position: An Approach to Future Prediction for Balasore Shoreline, Odisha, India, *Open Journal of Marine Science,* **5**, 13-25, (2015)
6. I. P. Joliffe, Coastal erosion, environmental – geological hazard, *Beaches and Coastal Geology,* 267–271, (1982)
7. E. Barthélemy, Nonlinear shallow water theories for coastal waves, *Surveys in Geophysics,* **25**, 315-337, (2004)
8. H. J. Verhagen, Aandrijfmechanisme voor Kuststromen, (1985)
9. FAO, The world's mangroves 1980-2005, (2007)
10. S. G. Thomas, A. Goudie, The Dictionary of Physical Geography, (2000)
11. H. Winterwerp, B. van Wesenbeeck, J. van Dalfsen, F. Tonneijck, A. Astra, S. Verschure, P. A. van Eijk, A sustainable solution for massive coastal erosion in Central Java, 45, (2014)
12. J. T. Houghten, Y. D. J. G. Ding, D. J. Griggs, M. Noguer, P. J. van der Linden, X. Dai, C. A. Johnson, Climate change 2001: the scientific basis, (2001)
13. J. C. Ellison, D. R. Stoddart, Mangrove ecosystem collapse during predicted sea-level rise: Holocene analogues and implications and implications, *Journal of Coastal research,* 151-165, (1991)
14. V. D. Prasita, Determination of shoreline changes from 2002 to 2014 in The Mangrove Conservation Areas of Pamurbaya using GIS, *Procedia Earth and Planetary Science,* **14**, 25-32, (2015)
15. W. Giesen, S. Wulffraat, M. Zieren, L. Scholten, Mangrove guidebook for Southeast Asia, *Mangrove guidebook for Southeast Asia,* (2007)
16. F. Blasco, P. Saenger, E. Janodet, Mangroves as indicators of coastal change, *Catena,* **27**, 167-178, (1996)
17. E. A. Himmelstoss, DSAS 4.0 Installation Instructions and User Guide, (2009)
18. J. L. Zichichi, A. Ergul, Digital Shoreline Analysis System (DSAS) version 4.0 – An ArcGIS extension for calculating shoreline change: U.S. Geological Survey Open-File Report 2008-1278

Analysis of static structural of ocean FARMITS offshore cage due to environmental load

Yani Purnawanti[1], *Nur* Syahroni[1] , and *Yeyes* Mulyadi[1]

[1]Ocean Engineering Department, Faculty of Marine Technology, Sepuluh Nopember Institute of Technology, Indonesia

Abstract. Consumption of fish in Indonesia continuously increase from 2000 to 2014. Particularly fish farming in Java from 2006-2014 is quite high, so there need to be adequate facilities for fish farming. Responding to these challenges, we developed Offshore Cage Ocean FARMITS technology specifically designed For the southern waters of Java Island. This paper discusses the static structure strength analysis of deformation and maximum yield strength that occurs on the offshore cage structure of Ocean FARMITS. The frame structure made of hollow pipe made from HDPE will be placed at 4,41m Wave Height, with a period of 19s. Numerical simulations are performed to obtain a motion response RAO and structural response due to random waves. This Structure is then analyzed under mooring conditions with the Catenary Mooring configuration. Rope stress that occur due to environmental loads will give tension stress to the structure, resulting in the offshore cage structure to occur maximum stress and deformation. The result of this research is to prove that the structure that is made by HDPE able to restrain the tension force of mooring line stress. This has been validated by the standard of Practical Aspects of Offshore Aquaculture System Design from Aquaculture America, so this structure can be said to be operating safely under Indonesian Irregular Wave.

1. Introduction

Consumption of fish in Indonesia continuously increase from 2000 to 2014. Particularly fish farming in Java from 2006-2014 is quite high, so there need to be adequate facilities for fish farming. Aquaculture technology began to develop since the 1970s, to date has grown rapidly in the food procurement sector with an increase of 32.4% in 2004. The sea-cage industry has existed in Norway, Chile, China, Spain, Australia, Japan, Indonesia, Scotland, Ireland, Canada, Greece and Faroe Island with various species cultivation[1].

Ocean FARMITS is a new innovation developed by Marine Reearch Center LPPM ITS Surabaya. This design is very suitable for Indonesian waters. In addition to the maximum cage volume, fabrication can also be done easily. For a mooring design of Ocean FARMITS Offshore Cage using catenary mooring system as in Figure 1.

This mooring system allows the movement of the structure to be rather large but can decrease mooring line tension. The combined mooring system of several cages used rectangular array as is moored to buoy, then buoy moored to anchor on the seabed. Such mooring systems are more effective than single mooring systems[2].

The study of parameter sensitivity on mooring line floating cage states that mooring loads are not sensitive to wave load variations for floating Cage structures with elastic floater[3]. But the effect of wave height when in irregular condition is more than in regular conditions [4]. The speed of ocean currents is also very influential on the floating cage stability. Further research is about the single dynamic behavior of point mooring semisubmersible fish cage with variations of current velocity[5].. The numerical analysis of mooring rope stresses and deformation of floating fish Cage structures was performed by Huang, et.al (2016). Based on the above reference we would like to do research on Offshore Cage Ocean FARMITS [6].

This paper aims to determine the effect of environmental condition on motion response and structural strength. Motion response due to environmental load causes tension on the mooring rope. The Strength of the structure can be seen through the magnitude of stress and deformation that occur due to the tension force of the mooring rope on the structure.

2. Description of Models

2.1 Detail Structure Offshore Cage Ocean FARMITS

Ocean FARMITS is a new innovation developed by Marine Research Center LPPM ITS Surabaya. This design is very suitable for Indonesian waters. In addition to the maximum cage volume, fabrication can also be done easily. Structural model of Ocean FARMITS Offshore cage is used to strength analysis with FEM simulation. This design uses HDPE PE 100 for material with yield strength 26 Mpa and tensile strength 31,7 Mpa [7]. Mooring lines on Offshore Cage Ocean FARMITS use nylon rope properties from a field handbook Aquaculture Operation in Floating HDPE cage 2015.

Offshore Cage Ocean FARMITS is designed for 6100 m3 cage volume. For the initial design use the dimension D=28m and H=10m as seen in Figure 1. Detailed size of the Offshore Cage Ocean FARMITS structure is shown in Table 1. The data in Table 1 refers to the reference "Aquaculture for Floating HDPE Cage"[8]. Details of parts of the structure in the table can be seen in the Figure 1.

Table 1. The Dimension of Offshore Cage Ocean FARMITS

Part	OD	Thickness	Quantity	Length
Pontoon Ring Outside	0,40 m	0,02 m	1	62,2 m
Sling	0,10 m	0,01 m	8	19,8 m
Horizontal Pontoon	0,20 m	0,01 m	8	4,95 m
Pontoon Ring Inside	0,28 m	0,02 m	1	31,1 m
Upper pontoon ring	0,25 m	0,01 m	1	15,5 m
Umbrella Frame	0,20 m	0,01 m	6	6 ,0 m

The detailed structure can be seen in figure 1. In Figure 1 it is seen that the structure of Ocean FARMITS has 3 main parts, there are the deck, upper deck, and underwater structure. At the underwater there is a damper to reduce the movement of the structure. There is also a sinker tube that serves as a ballast system. On the deck, there are several brackets as reinforcement. At the top of the deck, there is an umbrella frame to easily fish harvesting process.

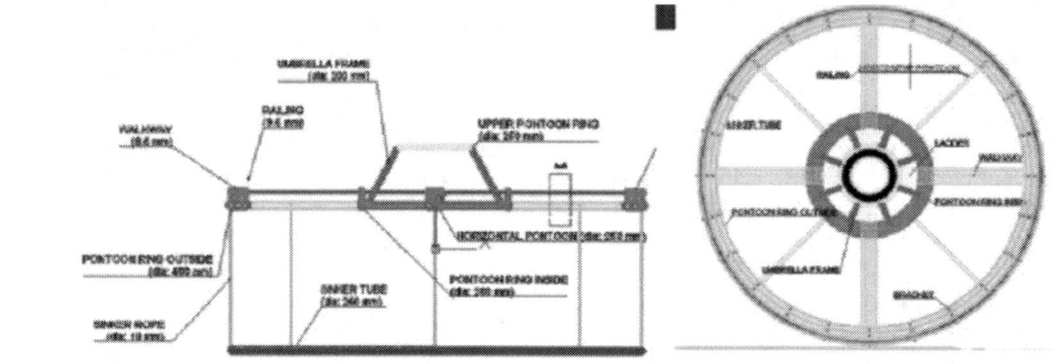

Fig. 1. Detail Structure Offshore Cage Ocean FARMITS

Environmental load (wind, wave, current) on the Offshore Cage Ocean FARMITS structure accumulates in the tension mooring line load. The size of the tension on the mooring line is the result of structural displacement due to the hydrodynamic load (surge, sway, heave, roll, pitch, yaw). In this structure the hydrodynamic load is analyzed under hurricane conditions, recording 100 year of wave, currents and wind data. For more details, the scheme in figure xx is a condition when the structure is exposed to the maximum tension load.

2.2 Hydrodynamics Model

Hydrodynamic modeling is performed to obtain the motion response of Ocean FARMITS structure on regular and irregular waves. Motions of structure on a regular wave can be shown through RAO (Response Amplitude Operator). RAO is obtained from the frequency domain numerical simulation. [9]

The floating cage under consideration consists of six degree of freedom modes, assuming that the oscillation movements are linear and harmonic, the coupling differential equation s can be written as follows equation 1:

$$\sum_{n=1}^{6}\left[\left(M_{jk}+A_{jk}\right)\ddot{\xi}_k+B_{jk}\dot{\xi}_k+C_{jk}\xi_k\right]=F_j e^{iwt}, j=1 \qquad (1)$$

where Mjk is the components of structure's mass matrix, Ajk, Bjk are matrix of added mass and damping coefficient, Cjk is restoring force coefficients, and Fj is amplitude of excitation force in complex quantities. F1, F2, dan F3 is the amplitude of excitation forces which result in surge, sway, and heave, while F4, F5, and F6 are the amplitude of excitation moments for roll, pitch, and yaw.

2.3 Structure Response due to Environmental Load

The structure response can be known through the multiplication of RAO squares with the wave spectrum. The response spectrum can be defined as the distribution of wave structure energy. It is formulated on equation 2 :

$$S_R=\left[RAO(\omega)\right]^2 S(\omega) \qquad (2)$$

where S_R spektrum respons, S(w) spektrum gelombang RAO (w) transfer function and w frekuensi gelombang. The wave spectrum equation used in this analysis is the JONSWAP spectrum equation because the parameters are included according to the characteristics of the closed seas or islands [10]. The JONSWAP spectrum equation is as follows equation 3:

$$S(\omega) = \alpha\, g^2 \omega^{-5} EXP\left[-1{,}25\left(\frac{\omega}{\omega_0}\right)^{-4}\right] \gamma^{EXP\left[\frac{-(\omega-\omega_0)^2}{2\tau^2\omega_0^2}\right]} \qquad (3)$$

where

α : 0,076 (X0)-0.22
X0 : gX/Uw2
X : fetch length
Uw : wind speed
α : 0.0081 if X not defined
ϒ : peak parameter, can be defined by equation 4:

$$\gamma = EXP\left[3{,}4843\left(1-0{,}1975\left(0{,}036-0{,}0056\frac{T_P}{\sqrt{H_S}}\right)\frac{T_P^{\,4}}{H_S^{\,2}}\right)\right] \qquad (4)$$

where
Tp : peak period of spectra
Hs : Significant wave height
τ : shape parameter (for ω ≤ ωo = 0,07 and ω ≥ ωo = 0,09)
ωo : $2\pi\left(\frac{g}{U_\omega}\right)(X_o)-0{,}33$

Numerical simulations are performed with free floating and moored conditions. It is as depicted in the Figure 2.

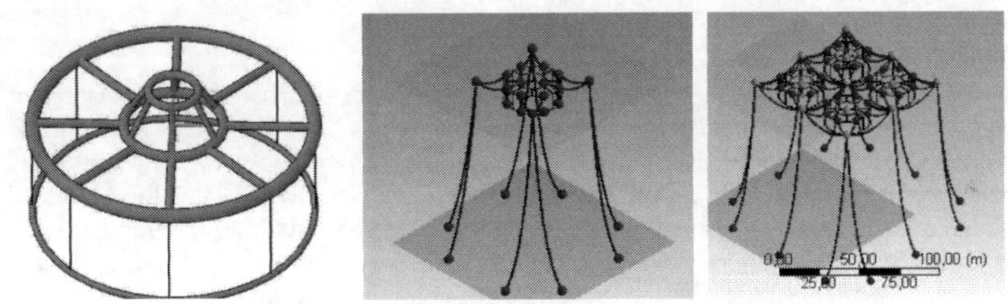

(a) (b) (c)

Fig. 2. (a) Surface Model (b) single moored condition (c) combined mooring condition

Figure 2.a is a surface model of Offshore Cage Ocean FARMITS for hydrodynamics diffraction analysis on free floating conditions. Figure 2.b is a surface model in moored condition, whereas in Figure 2.c is a combination mooring condition. This mooring configuration is more effective in its operation.

2.4 Finite Element Model

Finite element model is used in the calculation of mooring tension and structure stress of Ocean FARMITS. On the mooring line, each rope is divided into segments. Where there is a node on each end of the segment. In each segment rope is modeled its axial properties only. The properties of the mooring rope are mass, weight, buoyancy, cross section area, etc.The load is applied to nodes in the form of force and moment (weight, buoyancy, hydrodynamic drag, tension, etc.). Bending properties are facilitated by springs in segments[6]. The Equation 5 is for each segment nodal:

$$M(p.a) + C(p,v) + K(p) = F(p.v,t) \qquad (5)$$

where M(p,a) is the system inertia load, C(p,v) is the system damping load, K(p) is the system stiffness load, F(p,v,t) is the external load, t is the simulation time, p,v and a are the position, velocity, and acceleration.

FEM modeling on structures is required to obtain maximum stress and maximum deformation. In the picture xx meshing has been done on the structure to divide into several elements and nodal. In this case the element used is a solid element with the boundary condition shown in the Figure3b. In the Figure 3c section A to the ends are given fixed support while at the ends of element B is given a roll. C, D, E, F are the force inputs given from the maximum tension mooring line results. FEM modeling is only half done just to review the section affected by the maximum force.

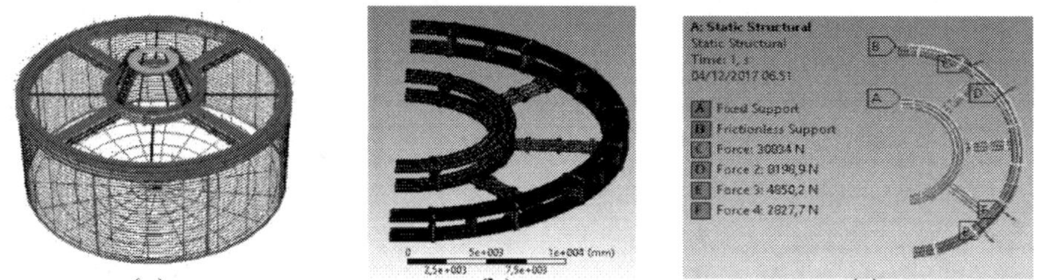

(a) (b) (c)

Fig. 3. (a) detail structure model (b) structure meshed (c) boundary condition

3. Environmental Data

The environmental data in Table 2 is the result from the processing of satellite images on the southern waters of Java Island precisely at the coordinates of Longitude 108.02 Latitude 7.88 with water depth 102 m. Environmental data such as wave height, wave period, and wind speed, are obtained from ECMWF satellite extraction (European Center for Medium-Range Weather Forecasts). Observations are made between 06:00 and 00:00. Depth data is obtained from GEBCO (General Bathymetric Chart of the Oceans). As for the current velocity obtained from OSCAR (Ocean Surface Current Analyses Real-time).

Table 2. Environmental Load on the southern waters of Java Island

Parameter		Unit	Storm Condition
Waves	Height (Hs)	m	4,41
	Period (Tp)	s	11,61
Wind	Velocity (Vw)	m/s	4,5
Current	Velocity (Vc)	m/s	0,3

Environmental loads in Table 2 are illustrated by JONSWAP spectrum theory become as in figure 4. In the Figure 4 the wave frequency is located at a frequency of 0.6 rad / s.

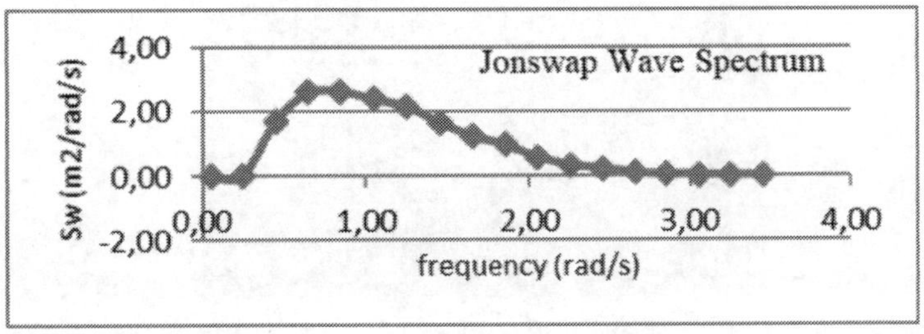

Fig. 4. Jonswap Wave Spectrum

4. Result and Discussion

4.1 Motion

Offshore Cage Ocean FARMITS oscillates in translational and rotational. Motions in this structure can be seen through RAO. The graph of RAO in Figure 5 shows the magnitude of movement of structures in the regular wave at each m of wave height. RAO is a transfer function for structural response due to random wave. The image is one of the largest RAO forms of rotational oscillation roll motion in free floating and moored conditions. At the time of free floating Figure 5(a) maximum roll reaches 7.6 deg/m at a frequency of 0.7 rad / s while at moored roll the maximum occurs at a frequency of 1 rad / s of 3.9 deg/s. This indicates that mooring condition can change the natural frequency of the structure shifts away from the wave natural frequency, due to the weight and stiffness of the mooring line. This is resulted in the magnitude of tension on the mooring line.

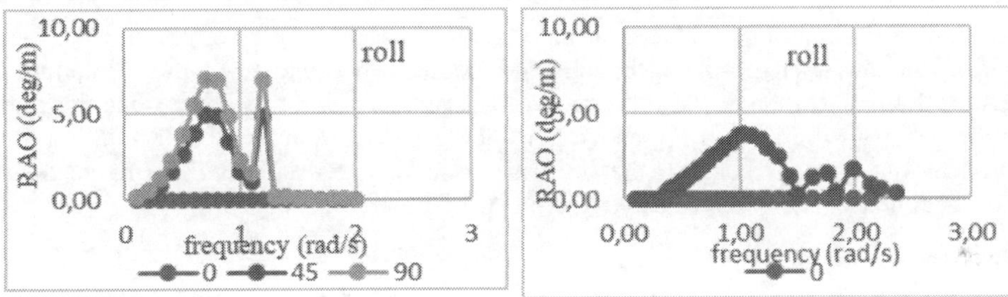

(a) (b)

Fig. 5. (a) Roll Motion Free Floating (b) Roll Motion Moored

4.2 Tension

Tension on the mooring line occurs because the lines hold the structural motions due to environmental load. The numerical simulation to get maximum tension is done with time domain analysis. This simulation can accommodate these non-linear factors, so the motion equation of Newton's second law is solved in time function or better known as time domain analysis. The equation is described as Equation 6:

$$[m + A(\omega)]\ddot{x} + C(\omega)\dot{x} + D_1\dot{x} + D_2 f(\dot{x}) + Kx = q_{WI} + q_{WA}^{1} + q_{WA}^{2} + q_{CU} + q_{ext} \quad (6)$$

Where q_{WI} wind drag force, q_{WA}^{1} first order wave load, q_{WA}^{2} second order wave load, q_{CU} current load, q_{xet} other external load.

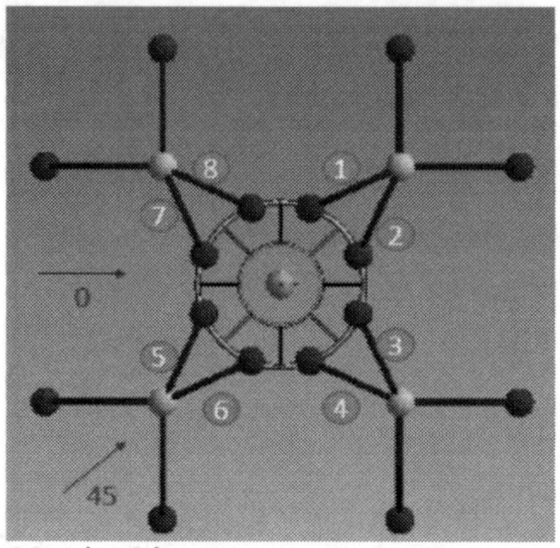

Fig. 6. Load Condition on the Mooring Line

Table 3. Maximum Tension on the Mooring Line

Dir.	Maximum Tension (N)								Time
	Line 1	Line 2	Line 3	Line 4	Line 5	Line 6	Line 7	Line 8	
0	9903	9783	8910	9031	38610	15100	38751	15231	26s
45	13225	13220	3536	9863	51403	51370	9764	3537	62s

Figure 6 is a loading condition on the mooring line. The direction of the environmental load comes from the direction of 0 degrees and 45 degress, because from this direction it produces the dominant response. Maximum tension is presented in the Table 3 occurs when the environmental load direction comes from 45 degrees. Maximum tension occurs on line 5 and line 6.This happens because the lines are loaded directly. This mooring tensions analysis was done by time domain simulation on storm condition.

4.3 Stress and Deformation

The tension that occurs on the mooring line leads to tension on the Offshore Cage Ocean FARMITS structure. Like the scheme in Figure 6 and Figure 3c, the mooring line is attached to the pontoon structure of Offshore cage Ocean FARMITS. So if the tension on a large mooring line, then the voltage that occurs in the structure is also large. Strength analysis on this structure is done to check the von misses stress that occurs in the structure. Von Misses Stress is derived from the Equation 7:

$$\sigma_{eq} = \frac{1}{2}\sqrt{(\sigma_x-\sigma_y)^2+(\sigma_y-\sigma_z)^2+(\sigma_z-\sigma_x)^2+6(\tau_{xy}^2+\tau_{yz}^2+\tau_{zx}^2)} \quad (7)$$

where

σ_{eq} *von mises stress*, σ_x normal stress x axis, σ_y normal stress y axis, σ_z normal stress z axis, τ_{xy} shear stress xy plane, τ_{yz} shear stress yz plane, τ_{zx} shear stress zx plane.

Maximum stress is obtained from the Finite Element Method, which before generating the stress of the model is done meshing to divide the segment and nodal. The meshing size is 0.5mm. This is the most optimum meshing size after meshing sensitivity analysis as in Figure7b. In the Figure 7a is shown one part of the structure that has been done meshing. Meshing is done by dividing critical and non critical areas. Critical area is an area that is exposed to direct force so that required a more dense meshing. For the critical meshing area use hexahedral meshing with optimum size 0.5 mm on each element. While for noncritical areas use triangular meshing with a larger size lead to decrease very long time for the simulation.

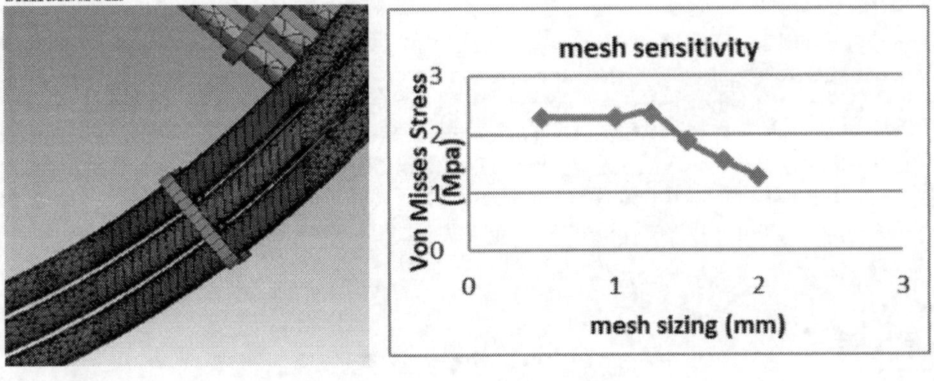

(a) (b)

Fig. 8. (a) Structural mesh (b) sensitivity analysis

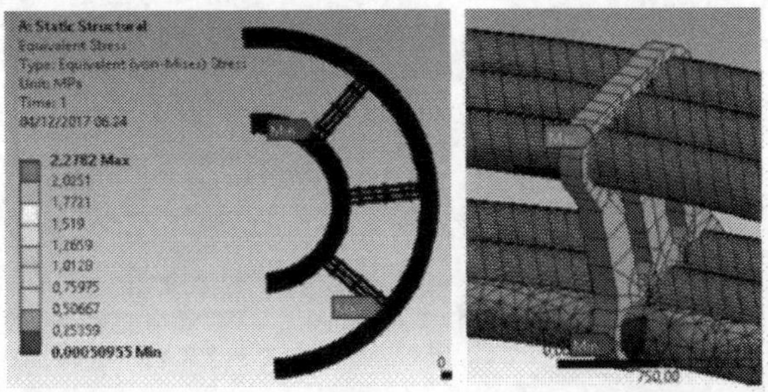

Fig. 9. Maximum Von Misses Stress

Figure 8 is the maximum stress result on the Offshore Cage Ocean FARMITS structure. Maximum tension occurs in the upper bracket area as it retains the deflection from the pontoon where the rope is tied. The mooring scenario is done on the upper pontoon, so the pontoon is deflected. Maximum von misses stress in this structure 2,28 Mpa. This value is less than 26 Mpa (Yield Strength HDPE) so this structure is safe to operate in this location even in storm conditions.

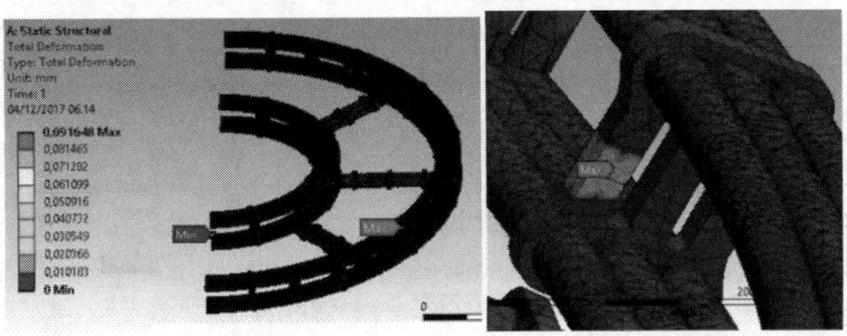

Fig. 10. Total Deformation

Figure 9 is the maximum deformation that occurs in the middle of the bracket. According to previous research about HDPE collar cage[6], the deformation for OD = 0.4m and thickness 20mm, the strain should be 0.4% -0.5%. So, for this case the maximum deformation should be less than 0.31 mm.

5. Conclusion

Environmental conditions greatly affect the strength of the structure. This is related to the motion of the structure generated, then the tension force on the mooring line to withstand movement, attributed to the upper pontoon of the structure. This gives the result that the effect of Environmental Load on Southern Waters of Java Island is Maximum tension is obtained when the structure is affected by 45 kN. Maximum Stress is located on the bracket of 2.28 MPa. This value is still far from the Yield Stress HDPE PE 100, so the structure is still safe in storm conditions. Maximum deformation is 1 mm located on the middle of bracket.

References

1. FAO. *The state of world fisheries and Aquaculture*. Rome: Food and Agriculture Organization of the United Nations; (2006) p. **162**.
2. Goudey, C.A. *Practical Aspects of Offshore Aquaculture System Design*. In Presentation at the World Aquaculture Society: Aquaculture America 2009. Seattle. America; (2009).
3. Kristiansen, T., Faltinsen,O.M. Experimental and numerical study of an Aquaculture net Cage. *Elsevier*, (2014) p.**1-26**.
4. Shainee, M.et.al., Investigation of Self-*Semisubmersible* SPM *Cage*s System in Random Waves. *Elsevier* (2013).
5. Decew, et.al. Modeling of dynamic behavior of a tunggal-point moored submersible. *ELsevier*, (2010) p.**38-45**.
6. Huang, X. et.al. Numerical Simulation Of Deformations And Forces Of A Floating Fish Cage Collar In Waves. *Elsevier*, (2016) p.**111-119**.
7. Vielma,J.,Kankainen,M. *Offshore Fish farming Technology in Baltic Sea Production Conditions*. Part-financed by the European Union Baltic Sea Region (2013).
8. Cardia, F., Lovatelli, A., *Aquaculture operations in floating HDPE Cages*. Rome : FAO and Ministry of Agriculture of the Kingdom of Saudi Arabia (2015).
9. Bhattacharyya. R., *Dynamics of Marine Vehicles*, New York: John Wiley & Sons Inc.(1978).
10. Chakrabarti, S.K., *Hydrodynamics of Offshore Structures*, USA: Computational (1987).

Submarine rudder stern-plane configuration for optimum manoeuvring

Ketut Suastika[1,*], *Putri* Virliani[2], and *Wasis* D. Aryawan[1]

[1]Department of Naval Architecture, ITS Surabaya, 60111 Surabaya, Indonesia
[2]LHI BPPT, 60111 Surabaya, Indonesia

Abstract. The purpose of the present study is to investigate submarine rudder stern-plane configuration to obtain optimum manoeuvring characteristics. Three configurations are investigated: +, x and y configuration. The equations of motions were derived with the external forces and moments consist of contributions due to the hydrostatics, hydrodynamics and propeller thrust. To obtain the hydrodynamic coefficients and the hydrodynamic derivatives, the lift and drag were calculated using computational fluid dynamics (CFD). The equations of motions were numerically integrated using Eulerian method to obtain the turning circle. Results of the calculations show that the x-configuration gives the smallest tactical diameter, transfer and advance, which is considered as the most optimum rudder stern-plane configuration.

1 Introduction

A submarine is expected to have a good manoeuvring performance for the achievement of her mission. One requirement is the ability to dive quickly to a certain depth. The other is the ability to manoeuvre extremely to avoid a mine or torpedo attack. The rudder stern-plane configuration particularly plays a key role in determining the submarine manoeuvring capability [1].

A midget submarine [2] is considered in the present study as illustrated in Fig. 1. The submerged displacement of a midget submarine is usually less than 150 ton. The principal particulars of the midget submarine are summarized in Table 1.

Earlier studies have considered similar object as considered in the present study. Erwandi et al. [3] studied the hydrodynamic forces working on the submarine while manoeuvring. Utina et al. [4] performed experiments on hydroplanes to study the hydrodynamic forces and moments for which a modified design of U-209 submarine was considered.

The present study pursues the earlier studies as reported in [3, 4]. The purpose of the present study is to investigate the manoeuvring performance of the submarine for which three different rudder stern-plane configurations are considered: +, x and y configuration, as shown in Fig. 2.

* Corresponding author: k_suastika@na.its.ac.id

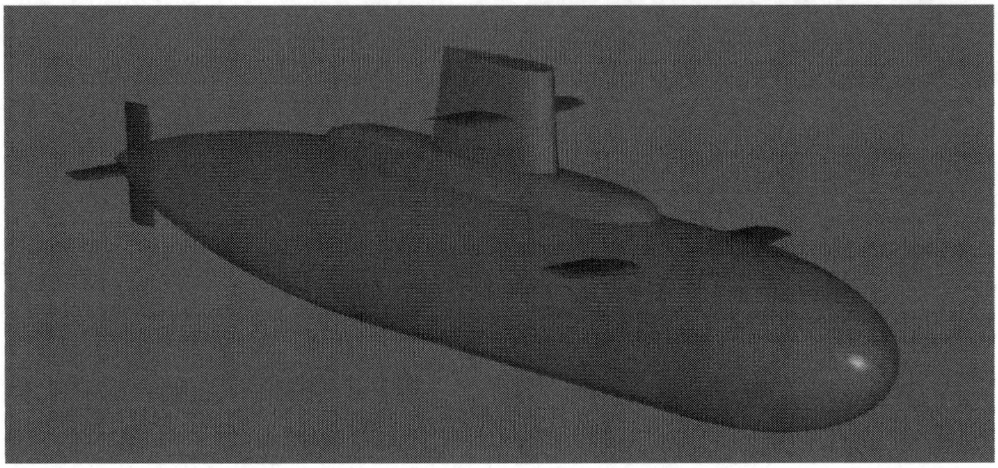

Fig 1. Midget submarine considered in the present study with + rudder stern-plane configuration.

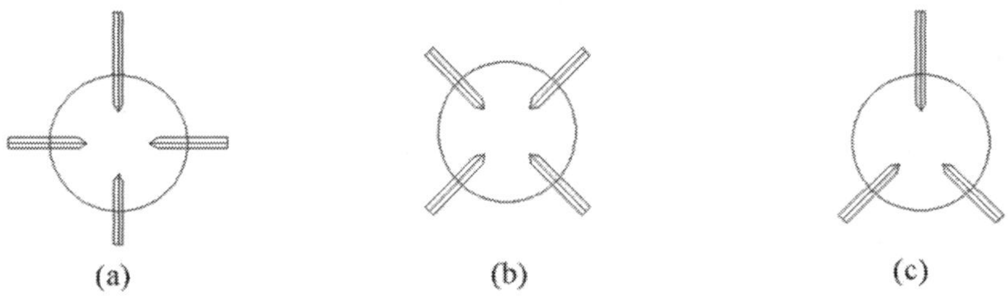

Fig. 2. Submarine rudder-stern plane configurations: +-stern (a), x-stern (b) and y-stern (c).

Table 1. Principal particulars of the midget submarine.

LOA	22.00 m
Diameter	3.00 m
Draft	2.60 m
Maximum speed	14 kn
Displacement (surfaced)	113.90 t
Displacement (submerged)	129.30 t
Volume of displacement (surfaced)	111.17 m^3
Volume of displacement (submerged)	126.14 m^3

2 Method

The equations of motions were derived with the external forces and moments consist of contributions due to the hydrostatics, hydrodynamics and propeller thrust. Subsequently, the equations of motions are numerically integrated to obtain the turning circles of the submarine for the different variations of rudder stern-plane configurations.

2.1 Equations of motions

The submarine motion consists of translation respectively in the x, y and z direction (surge, sway, heave) and rotation respectively about the x, y and z axis (roll, pitch, yaw/drift), which make a 6-degrees of freedom (DOF) motion. The notations used for the motions, forces and moments, together with the local coordinate system moving with the boat are shown in Fig. 3. The origin O of the local coordinate system (x, y, z) is located at the centre of buoyancy of the boat (for a fully submerged boat, the centre of buoyancy is fixed).

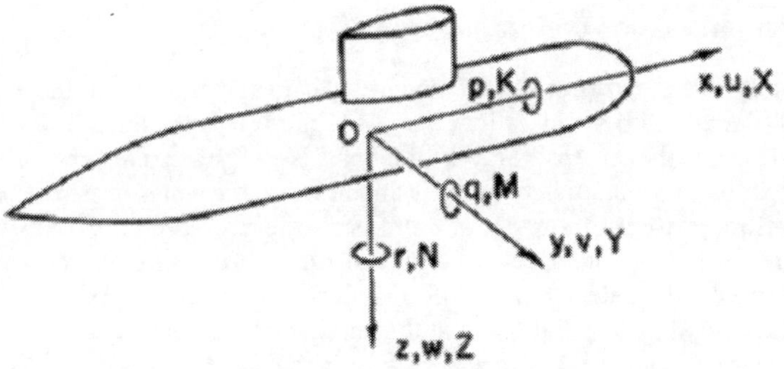

Fig. 3. Local coordinate system and notations for the motions [5].

Using the notations shown in Fig. 3, the equations of motions can be represented as follows:

$$\begin{bmatrix} m-X_{\dot{u}} & 0 & 0 & 0 & mz_G & -my_G \\ 0 & m-Y_{\dot{v}} & 0 & -mz_G & 0 & mx_G-Y_{\dot{r}} \\ 0 & 0 & m-Z_{\dot{w}} & my_G & -mx_G-Z_{\dot{q}} & 0 \\ 0 & -mz_G & my_G & I_x-K_{\dot{p}} & 0 & 0 \\ mz_G & 0 & -mx_G-M_{\dot{w}} & 0 & I_y-M_{\dot{q}} & 0 \\ -my_G & -mz_G-N_{\dot{v}} & 0 & 0 & 0 & I_z-N_{\dot{r}} \end{bmatrix} \cdot \begin{bmatrix} \dot{u} \\ \dot{v} \\ \dot{w} \\ \dot{p} \\ \dot{q} \\ \dot{r} \end{bmatrix} = \begin{bmatrix} \Sigma X_{ext} \\ \Sigma Y_{ext} \\ \Sigma Z_{ext} \\ \Sigma K_{ext} \\ \Sigma M_{ext} \\ \Sigma N_{ext} \end{bmatrix} \quad (1)$$

where m is the mass of the boat, I_x, I_y and I_z are the moment inertia's about the x, y and z axis, respectively, $X_{\dot{u}}$, $Y_{\dot{v}}$, $Z_{\dot{w}}$, $K_{\dot{p}}$, $M_{\dot{q}}$, $N_{\dot{r}}$, $Y_{\dot{r}}$, $Z_{\dot{q}}$, $M_{\dot{w}}$ and $N_{\dot{v}}$ are added masses. The external forces and moments on the right-hand side of eq. (1) consist of contributions due to the hydrostatics, hydrodynamics and propeller thrust.

The hydrostatic forces consist of weight and buoyancy, which make an equilibrium. The buoyancy is calculated using the software Maxsurf. The buoyancy force is equal to the weight of the boat (1268379.29, 1268992.70 and 1267846.36 N for the +, x and y-stern, respectively).

The hydrodynamic forces consist of added masses, drag and lift. The longitudinal and transversal added masses are calculated using the strip theory [6]. The rolling added mass is calculated using the formula given in [7]. The drag and lift are calculated using computational fluid dynamics [8]. The CFD results for the drag and lift are used to calculate the hydrodynamic coefficients.

The propeller thrust is calculated using an empirical formula given in [9].

The hydrodynamic derivatives required for the calculation of the turning circle are calculated using the formula given in [10, 11].

Results of the calculations of the added masses, drag, lift, thrust and hydrodynamic derivatives described above can be found in [12].

2.2 CFD calculations for the drag and lift

CFD simulations were performed to calculate the drag and lift of the submarine. The computational domain is defined as follows [13]. The inlet located at 1.5 L upstream from the model, where L is the length overall of the boat, is defined as the flow velocity equalling the boat's speed (in the simulation the boat is at rest while the water flows). The outlet located at 3.0 L downstream from the model is defined as the pressure equalling the undisturbed hydrostatic pressure. The boundary conditions at the bottom wall (9.0 D below the model, where D is the boat's diameter), top wall (9.0 D above the model) and side walls (9.0 D aside the model) are defined as free slip while at the surface of the boat as no-slip.

The turbulence model used is the shear-stress transport model [14]. The model gives accurate predictions of the onset and the amount of flow separation [15].

Grid independence tests were performed to ensure that the simulation results do not depend on the number of cells (elements) used in the calculations. The grid independence criterion is defined as the difference between two subsequently calculated drags is less than 2% where the number of cells used in the latter simulation is approximately twice the number of cells used in the former simulation [16, 17].

As criterion for the convergence of the numerical solutions, the root mean squared (rms) error criterion is used with residual target value of 10^{-5}.

The simulations were run in 64-bit OS stand-alone computer with Intel®Core i7-4770 CPU@3.40 GHz processor and 8.00 GB installed memory (RAM).

The drag calculated from the CFD is verified using experimental data obtained from towing tests [3, 4]. Figure 4 shows a comparison between the drag calculated from CFD and experimental data for the + rudder stern-plane configuration. Figure 4 shows that the CFD underestimates slightly the experimental data. The mean absolute percentage error is 2.8%, indicating a good agreement between the CFD results and the experimental data.

After verifying the CFD results with the available experimental data, the drag and lift coefficients for each rudder stern-plane configuration are calculated for boat's speed of 14 knots (maximum target speed) with varying rudder angle (δr) and drift/yaw angle (β). The rudder angle δr is varied from 0° to 35° with interval of 5° and the drift angle β is varied from 0° to 20° with interval also of 5°, as shown in Table 2.

Table 2. Variations of drift angle β and rudder angle δr in the calculation of drag and lift using CFD.

β [°]	δr [°]
0	0, 5, 10, 15, 20, 25, 30, 35
5	0, 5, 10, 15, 20, 25, 30, 35
10	0, 5, 10, 15, 20, 25, 30, 35
15	0, 5, 10, 15, 20, 25, 30, 35
20	0, 5, 10, 15, 20, 25, 30, 35

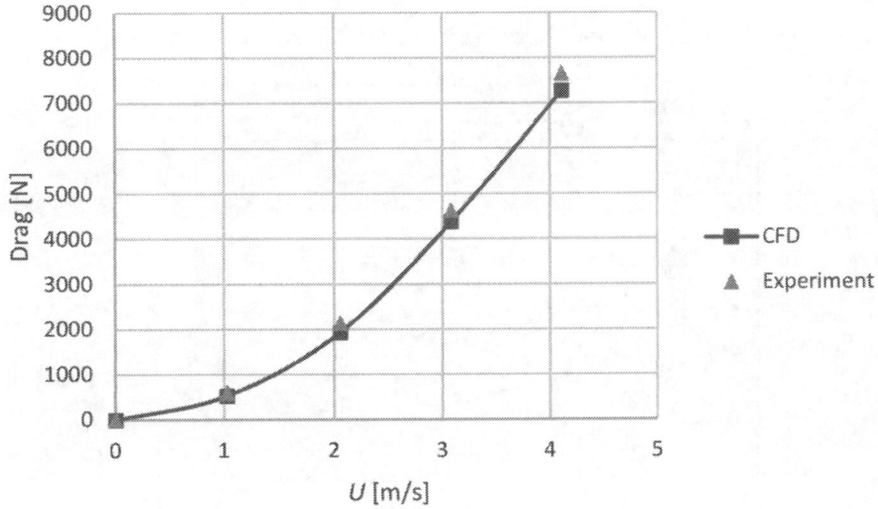

Fig. 4. Drag calculated from CFD compared with experimental data for the + rudder stern-plane configuration.

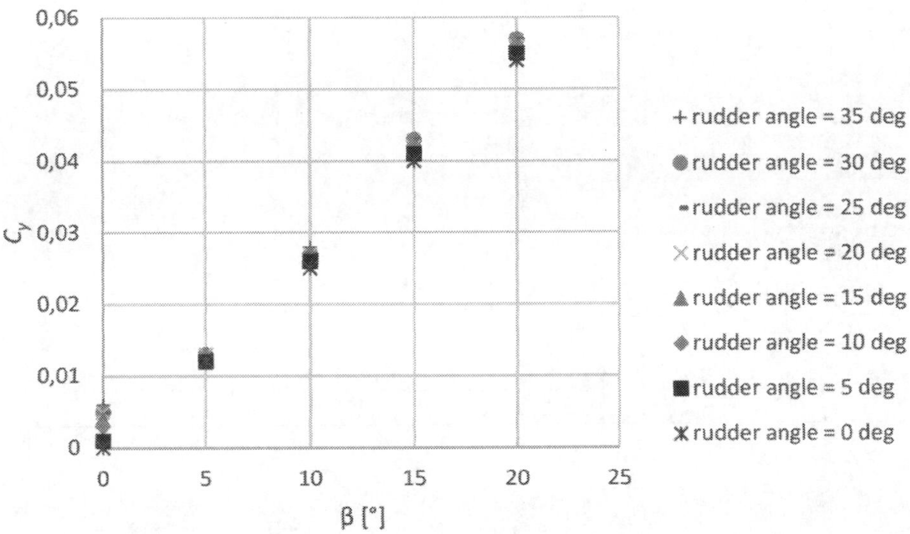

Fig. 5. Lateral lift coefficient C_y for the + rudder stern-plane configuration as function of drift angle β for varying rudder angle δr.

As an example of the calculation results, Fig. 5 shows the lateral lift coefficient (C_y) as function of yaw/drift angle (β) for varying rudder angle (δr) for the + rudder stern-plane configuration. As shown in Fig. 5, the lateral lift coefficient (C_y) increases with increasing drift angle (β). Furthermore, for a given drift angle, the lift coefficient C_y increases with increasing rudder angle.

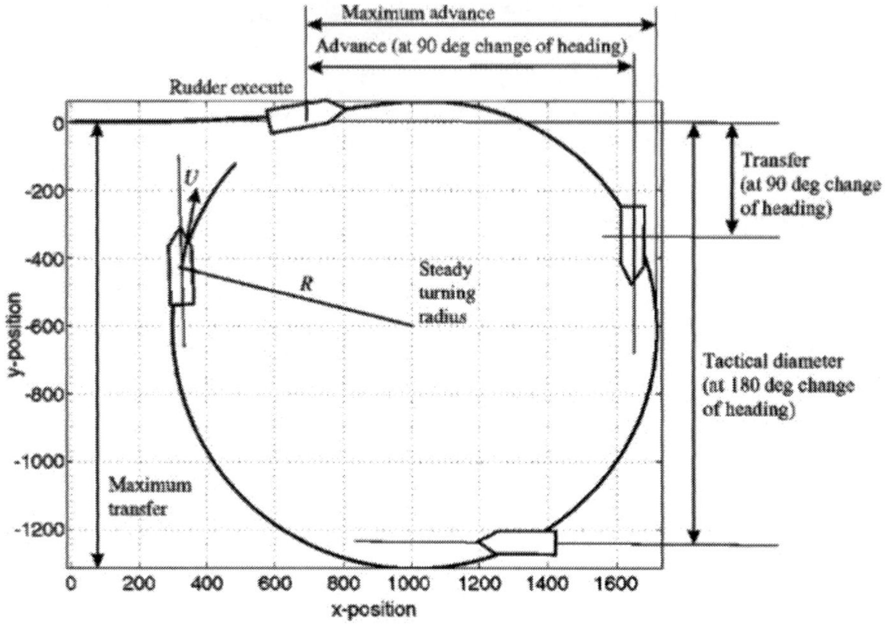

(a) Fossen [9]

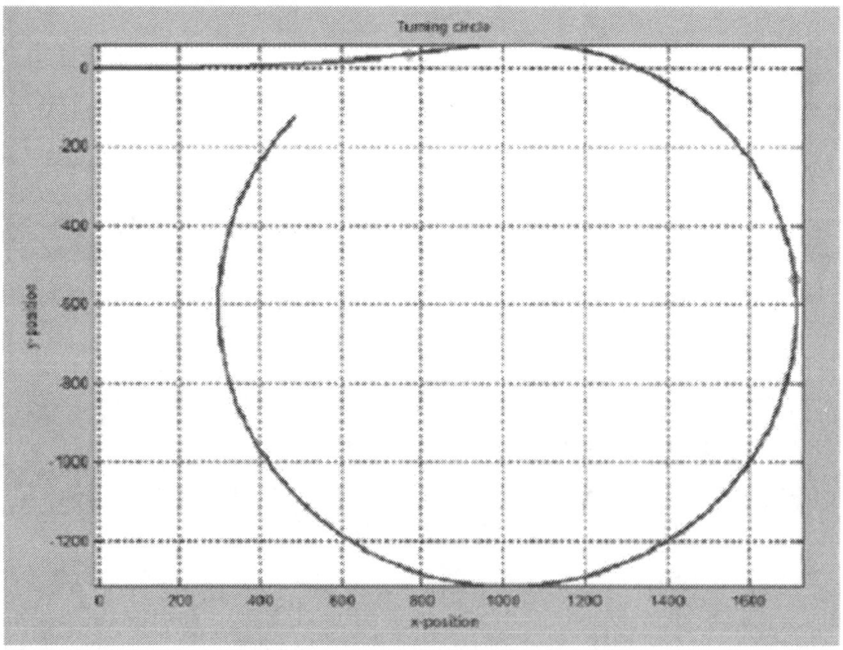

(b) Present study

Fig. 6. Comparison between turning circle calculated by Fossen [9] (a) and that obtained from the present study (b).

To obtain the turning circle [18] for a given rudder-stern plane configuration and rudder angle as well as yaw angle, the equations of motions (1) were integrated numerically using in Matlab implemented Eulerian method. The results were verified using those reported in [9]. Figure 6 shows a comparison between the turning circle reported in [9] and that obtained from the present study. The calculation results are summarized in Table 3. Table 3 shows that there are no differences between the results reported in [9] and the present results.

Table 3. Comparison between turning circle parameters calculated by Fossen [9] and those obtained from the present study.

	Fossen [9]	Present study
Rudder execute (x coordinate)	769 m	769 m
Steady turning radius	711 m	711 m
Maximum transfer	1315 m	1315 m
Maximum advance	947 m	947 m
Transfer at 90 degrees heading	534 m	534 m
Advance at 90 degrees heading	943 m	943 m
Tactical diameter at 180 degrees heading	1311 m	1311 m

3 Results and Discussion

The turning circle for each combination of rudder angle (δr) and drift/yaw angle (β) as shown in Table 2 can be plotted and the tactical diameter, transfer, advance etc. can be calculated. For example, Fig. 7 shows results of the turning circles with rudder angle of 10° and yaw angle of 0° for the +, x and y configurations projected in the (x_{pos}, y_{pos}) plane. Figure 7 clearly shows that the x-configuration gives the smallest turning circle radius.

In addition, Figs. 8, 9 and 10 show the turning circles in the (x_{pos}, y_{pos}, z_{pos}) space with the same rudder angle and drift angle as shown in Fig. 7 (δr = 10° and β = 0°) for the +, x and y configuration, respectively. Figure 8 shows that for the + configuration the turning circle remains in the same horizontal plane, as expected, because the vertical lift coefficient C_z is zero. On the contrary, the vertical lift coefficients of the x and y configurations are non-zero. Therefore, as shown in Fig. 9 (the x configuration), the submarine dives to a depth of 4 m below the initial position and then comes upward while circling to a position 2 m above the initial position. Similarly, for the y configuration (Fig. 10), the submarine dives 10 m below the initial position and then comes upward while circling until 5 m above the initial position.

Table 4 summarizes the results of tactical diameter for the case β = 0°. Table 4 shows that for all rudder stern-plane configurations, the larger the rudder angle, the smaller the tactical diameter, as expected. Furthermore, for all rudder angles, the x configuration gives the smallest tactical diameter (compared to the + and y configurations). Similar observations are also applied for the transfer, advance and radius of the turning circle.

Results for the turning circle are compared with the criteria given by the IMO resolution MSC 137 (76) for surface ship because there are still no corresponding criteria for submarine. Results of calculations show that to comply with the above IMO criteria, the smallest rudder angle is 25°, 10° and 20° for the +, x and y configuration, respectively.

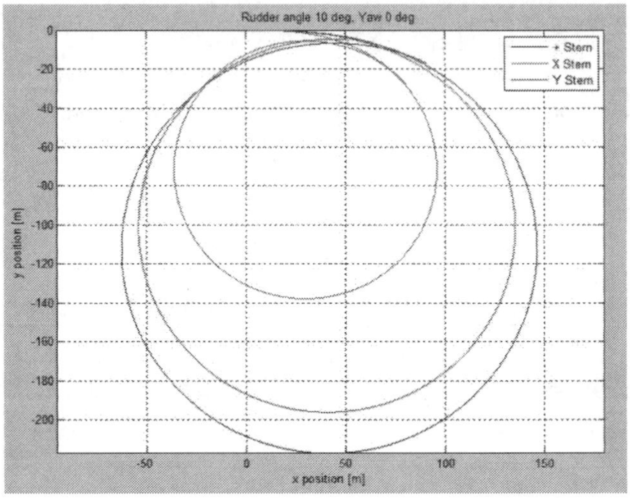

Fig. 7. Turning circles for rudder angle δr = 10°, yaw β = 0°.

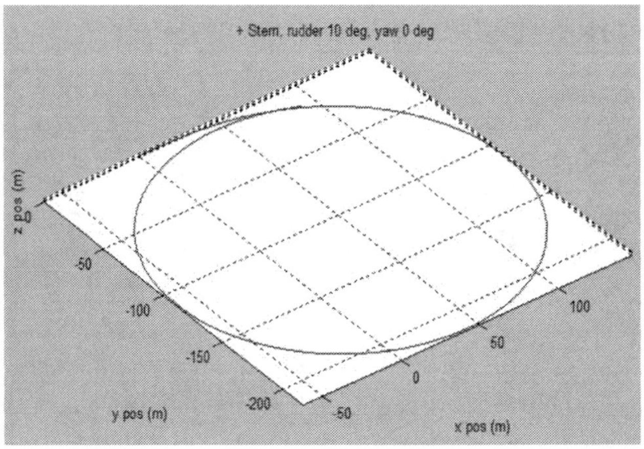

Fig. 8. Turning circle (x_{pos}, y_{pos}, z_{pos}) of the submarine for the + rudder stern-plane configuration with rudder angle δr = 10° and yaw angle β = 0°. In this case, the resulting z_{pos} = 0.

Table 4. Tactical diameter as function of rudder angle for different rudder stern-plane configurations (β = 0°).

Ruder angle [°]	Tactical diameter [m]		
	+-stern	x-stern	y-stern
5	1210	810	986
10	217	138	196
15	146	69	113
20	118	62	75
25	106	44	59
30	82	43	52
35	62	32	41

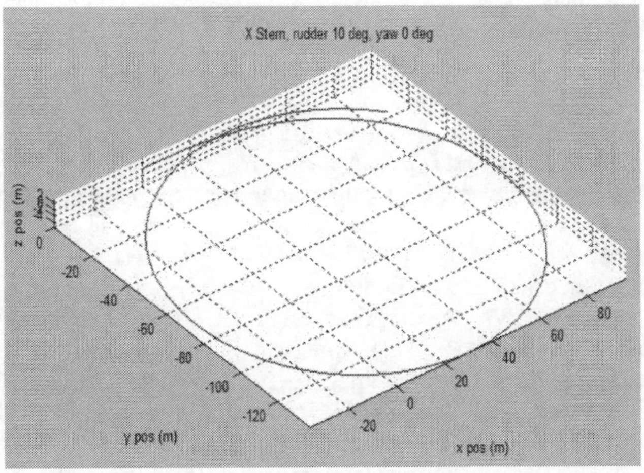

Fig. 9. Turning circle (x_{pos}, y_{pos}, z_{pos}) of the submarine for the x rudder stern-plane configuration with rudder angle δr = 10° and yaw angle β = 0°.

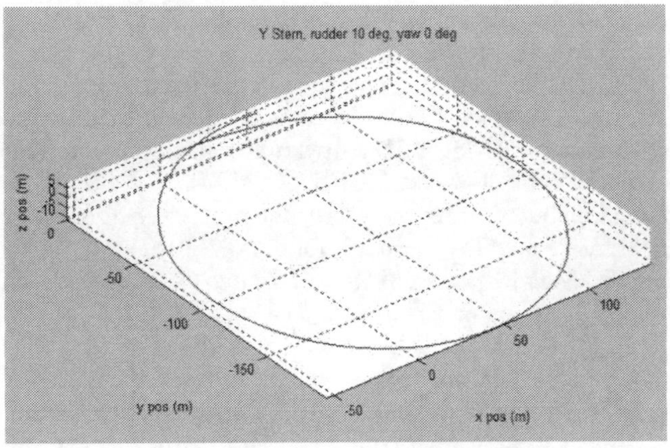

Fig. 10. Turning circle (x_{pos}, y_{pos}, z_{pos}) of the submarine for the y rudder stern-plane configuration with rudder angle δr = 10° and yaw angle β = 0°.

4 Conclusion

The rudder stern-plane configuration of a submarine plays a key role in determining the submarine manoeuvring performance. In the present study, three rudder stern-plane configurations (+, x and y configuration) were investigated to obtain optimum manoeuvring characteristics of a midget submarine. For all configurations, the larger the rudder angle, the smaller the tactical diameter, transfer and advance. For the + configuration, the smallest rudder angle to result in the tactical diameter, radius and advance that comply with the IMO resolution MSC 137 (76) is 25°, while for the x-configuration 10° and for the y-configuration 20°. The smallest tactical diameter, transfer and advance were obtained by using the x-configuration, which is considered as the most optimum rudder-stern plane configuration.

Putri Virliani thanks the Ministry of Research, Technology and Higher Education of the Republic of Indonesia for the scholarship Pascasarjana Dalam Negeri 2015-2017.

References

[1] R. Burcher, L. Rydill, *Concepts in submarine design*, Cambridge Univ. Press, (1994)
[2] M. Moonesun, P. Charmdooz, General Arrangement and naval architectural aspects in midget submarines, 4th Int'l Conf. Underwater Sys. Tech.: Theory Appl., Shah Alam, Malaysia (2012)
[3] Erwandi, M. Jefferson, C. S. Jati, D. Wahyudi, Nurwidhi, Desain dan uji hidrodinamika kapal selam mini berbobot 133 ton, Prosiding InSINas, pp. 1-6 (2012)
[4] M. R. Utina, A. Syafiul, B. Ali, Numerical and experiment investigation of lift performance over hydroplane of submarine, *J. Subsea Offshore Sci. Engrg.*, **5** (2016)
[5] M. Mackay, Wind tunnel experiments with a submarine afterbody model, Tech. Memorandum, Defence Res. Development, Canada (2003)
[6] T. I. Fossen, *Guidance and control of ocean vehicles*, John Wiley & Sons Ltd., New York (1994)
[7] R. D. Blevins, *Formulas for natural frequency and mode shape*, Krieger, Florida (1979)
[8] Toxopeus, Practical application of viscous-flow calculations for the simulation of manoeuvring ships, PhD tesis, Maritime Res. Inst., Netherlands (2011)
[9] T. I. Fossen, *Hand book of marine craft hydrodynamics and motion control*, Norwegian Univ. Sci. Tech., Trondheim, Norway (2011)
[10] Feldman, DTNSRDC revised standard submarine equations of motion, Tech. Rep. Carderock Div. Naval Surface Warfare Center, Maryland, USA (1979)
[11] H. Zhang, Y. Xu, H. Cai, Using CFD software to calculate hydrodynamic coefficients, *J. Marine Sci. Appl.*, **9** pp. 149-155 (2010)
[12] P. Virliani, Pengaruh bentuk after body terhadap maneuvering kapal selam mini, M.Sc.-thesis, Fac. Marine Tech., ITS Surabaya, Indonesia (2017).
[13] H. K. Versteeg, W. Malalasekera, *An introduction to computational fluid dynamics: the finite volume method*, Longman Scientific, Harlow, UK (2007)
[14] F. R. Menter, Two-equation eddy-viscosity turbulence models for engineering applications, *AIAA J.*, **32**(8), pp. 1598–1605 (1994)
[15] J. E. Bardina, P. G. Huang, T. J. Coakley, Turbulence modeling validation, testing, and development, NASA Tech. Memorandum 110446, Ames Res. Center, Moffett Field, California, USA (1997)
[16] J. D. Anderson Jr., *Computational fluid dynamics: the basics with applications*, McGraw-Hill, Inc., New York (1995)
[17] W. L. Oberkamp, T. G. Trucano, Verification and validation in computational fluid dynamics, *Prog. Aerospace Sci.*, **38** pp. 209-272, Pergamon Elsevier (2002)
[18] M. Renilson, *Submarine hydrodynamics*, Springer Briefs Appl. Sci. Tech. (2015)

Identification of factors influence to complétion of adoption procès of ecolabel in fisheries product

Ratna Purwaningsih[1,*], *Aries* Susanty[2], *Amru* Khaifa Wafa[3], *Ary* Arvianto[4], and *Zulfaidah* Ariany[5]

[1,2,3,4] Industrial Engineering Department, Engineering Faculty, Diponegoro University Jl. Prof. Soedharto, SH, Tembalang, Semarang, Central, Java, Indonesia 50275
[5]Naval Architecture and shipbuilding Department, School of Vocational, Diponegoro University Jl. Prof. Soedharto, SH, Kampus Undip Tembalang, Semarang, Central Java, Indonesia 50275

Abstract. This study aims to determine what factors affect consumers in the process of adoption of eco-label on fishery products and provide program recommendations to improve eco-label adoption. The study used primary data through the questionnaire in a Likert scale 1-5 to 200 respondents in Semarang City, Central Java Indonesia. The process of adoption of eco-label is divided into two phases: initiation of adoption process and completion of the adoption process, the discussion focus on completion phase. Hypothesis testing is done by multiple linear regression analysis. The results indicate that in the phase of completion of the adoption process, consumer buying intention and consumer experience is moderator variable that has the greatest influence to accelerate the completion of the adoption process.

1 Introduction

Since 2002, 72% of the world's marine fish stocks were being harvested faster than they could reproduce. Fishing activities have various negative impacts on marine ecosystems. The greatest concern is the rapid depletion of fish population due to extensive commercial fishing. Overfishing occurs when fish are caught faster than they can reproduce, and for many scientists, it has become one of the greatest impacts of human activity on oceans. Overfishing increases the vulnerability of ocean ecosystems and may contribute to the decline of other marine species [1]. Research results from the Indonesian Marine Fishery Research Center regarding the estimation of fishery status stated that 6 out of 11 WPP (Fishery Management Areas) of Indonesia showed clear symptoms of overfishing [2].

Eco-labels can be a solution to prevent overfishing [3]. Product certification and eco-labeling are tools that can be used to support fisheries management. The goal of the eco-labeling program is to create market-based incentives for better management of fisheries by creating consumer demand for seafood products from well-managed stocks. Eco-labels are seals of approval given to products that are deemed to have fewer impacts on the

* Corresponding author: ratna.tiundip@gmail.com

environment than functionally or competitively similar products [4]. Ecolabelling is only one type of environmental performance labeling and refers specifically to the provision of information to consumers about the relative environmental quality of a product. There are many different environmental performance labels and declarations being used or contemplated around the world [5].

Then, eco-labeling is a way to increase fish stocks from overexploited areas by providing market incentives for producers at premium prices on eco-labeled products [6]. Because of the voluntary nature of the eco-label implementation, producers need high market incentive prospects. Thus, consumers have an important role in determining the success of eco-labeling. Individual consumption has a major impact on the environment. Some experts estimate that the magnitude of impact can reach 30% to 40% [7]. Studies of consumer perceptions of eco-label are important to do, although they are often confused by the terminology used on the label.

There are two phases of consumer process in adopting eco-label, first is an initiation of the adoption process and second is a completion of the adoption process. Each phase is influenced by several variables either directly or moderated variables. Thogersen, pointing out that in initiating the process of eco-label adoption in fishery products, consumers are significantly influenced by several factors: environmental stewardship, knowledge of sustainable fisheries, subjective knowledge of sustainable fisheries, and innovation [8]. High effort makes it difficult to make eco-label work effectively to contribute to the environment and producers. One of the moderator variables is consumer confidence, experience and buying desire. Then another research says that there is an important factor that also affects consumer to buy eco-labeled products on an ongoing basis environmental campaign [9].

Knowing the key factors affecting consumers in adopting eco-labels on fishery products can provide practical implications for encouraging companies to apply eco-labels that can later be used to create good marketing strategies that effectively impact companies and the environment. Based on the description above it will be conducted research to analyze the factors that affect consumers in adopting eco-label on fishery products.

2 Literature review

Eco-label comes from the word eco which means the environment and the label which means a mark on the product that differentiates it from other products. Eco-labels help consumers to choose environmentally friendly products as well as serve as a tool for manufacturers to inform consumers that the products they produce are environmentally friendly [10]. Eco-labels are provided through an independent third party certification process to assess that a product is manufactured with due regard to environmental preservation principles. Referring to the GATT (General agreement on tariff and trade), the eco-label is based on non-discrimination and on a voluntary basis. Voluntary basis means that the certification system works on the basis of market incentives. Manufacturers can see market incentives as willing to pay more (WTP) for labeled products or opportunities to develop new markets.

Adoption is the decision to fully use new ideas as the best way to act. Decision innovation is a mental process since someone knows the existence of innovation to take the decision to accept or reject it and then confirm it. Manifestations of this form of adoption can be seen or observed in the form of behavior, methods, and equipment and technology used in communication activities. Thogersen suggests that the adoption process on eco-label consists of two stages. The first stage is the initiation of the adoption process whereby consumers have begun to accept the concept and understand the purpose of the concept. The second stage is the completion of the adoption process whereby consumers will begin to try and continuously adopted.

Discussion in this paper is only on completion of the adoption process. Initiation of the adoption process already done and hypothesis test result stated that hypothesis testing H1 ($\beta = 0.806$; $p>0.05$) accepted. Chekima et all [9] adds a variable moderates consumers to continuously purchase eco-label products, it is environmental campaigns, as illustrated in the following model in figure 1.

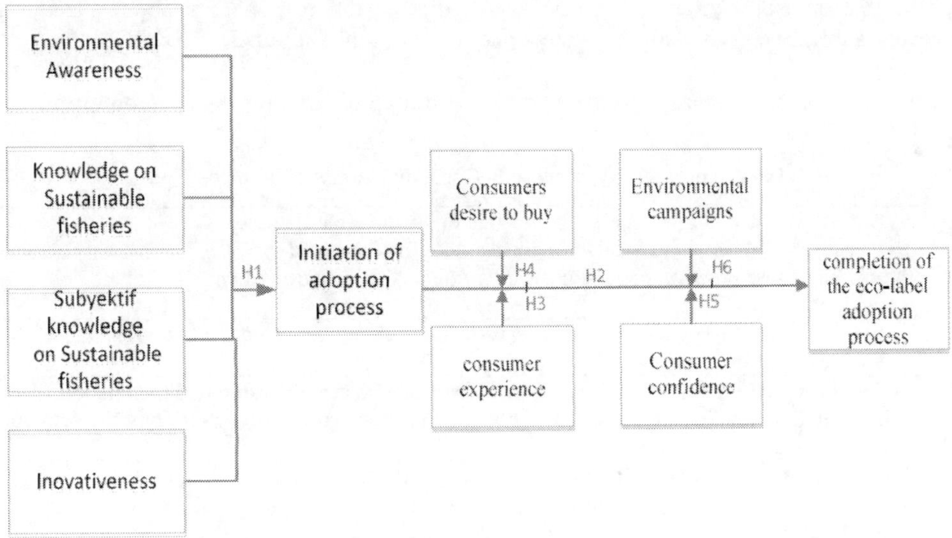

Fig. 1. Model of completion of the adoption process

Figure 1 describes that Initiation of adoption process influence by Environmental Awareness, Knowledge on Sustainable fisheries, Consumer subjective knowledge on Sustainable fisheries, and Innovativeness. The completion of the adoption process is a process moderated by four factors. The first factor is the desire to buy or effective response where consumers like and want to make a decision to buy eco-label fishery products. The second factor is the experience or process that brings someone to a higher behavioral pattern, includes purchasing eco-labeled products in the past. The third factor is consumer confidence or consumer perception to National Standardization Agency as the government institution that makes regulation of environmentally friendly products in Indonesia. The fourth factor is environmental campaigns which provide by society and government.

3 Research Methods

3.1 Research Sample and Questionnaire

In this study first conducted a preliminary study by distributing questionnaires to the people of Semarang city. The study used regression analysis, it was suggested that 15 - 20 observations per independent variable. Numbers of independent variables is 9 variable, the minimum number of respondents is 200 respondents. Respondents are the people of Semarang city who had ever bought any brand fishery products.

There are two parts of the questionnaire used in this research. Section 1 contains general data of respondents and questionnaire instructions. Section 2 contains assessments of six research variables: initiation of the eco-label adoption process, consumer buying desire, consumer trust, environmental campaigns, initiation of the adoption process, and the process of completing the adoption.

3.2 Research Hypothesis

The model of consumer eco-label adoption process gives a brief overview of factors affecting completion adoption process. According to the model, the research has six hypothesis listed in table 1. The first study was already done and proven the hypothesis one was accepted. Then, the result and discussion only focus on others hypothesis. The results showed that environmental awareness, knowledge of the respondents about sustainable fisheries, subjective knowledge of respondents and innovativeness proved to have the correlation with the initiation of the adoption process. The test results also state that each significantly positive effect on consumers when initiating the process of adoption of eco-label [11].

Table 1. Research Hypothesis on Consumer completion of ecolabel

H1:	The Environmental Awareness, Knowledge on Sustainable fisheries, Consumer subjective knowledge on Sustainable fisheries, and Innovativeness positively and significantly affects the Initiation of an adoption process
H2 :	The initiation of the eco-label adoption process positively and significantly affects the completion of the eco-label adoption process
H3:	The consumer experience becomes a moderator that positively influences the relationship between the initiation of the eco-label adoption process with the completion of the eco-label adoption process
H4:	Consumers desire to buy is a moderator that positively influences the improvement of the relationship between the initiation of the eco-label adoption process and the completion of the eco-label adoption process
H5:	Consumer trust is a moderator that positively influences improving the relationship between the initiation of the eco-label adoption process with the completion of the eco-label adoption process
H6:	The environmental campaign is a moderator that positively influenced the improvement of the relationship between the initiation of the eco-label adoption process and the completion of the eco-label adoption process

3.3 Data Processing

Data analysis was done by using compound linear regression. Linear regression is the statistical method used to form models relationship between dependent variable (Y) with one or more independent variables predictor (X). A variable moderator can strengthen or weaken the relationship between independent and dependent variables. Testing the statistics on moderator variables usually, use Moderated Regression Analysis (RAM). Therefore, linear regression analysis is used to measure influence between independent variables (X) that exist in the process of adoption of eco-label (Y).

4 Result and Discussion

Respondents were 87 men and 113 women. The majority of respondent was between 20 and 40 years old. A total of 94 respondents educated bachelor degree and 63 have high school education. Respondents' income spread almost equally between 1 million to 7 million per month, as many as 34 respondents earned more than 7 million per month.

4.1 Test of Validity and Reliability

Test the validity and reliability of the questionnaire are done through a pilot study by distributing questionnaires to 40 respondents those consumers of fishery products in the city of Semarang. The reliability test was done with a total of 30 respondents and use a significance level of 5%. Test obtained r table value of 0,361. The value of r arithmetic obtained at the value of Pearson product moment. Reliability test questionnaire is done based on Cronbach's Alpha value. If the value of Cronbach's Alpha is greater than 0.6 then the relevant variable is said to be reliable. The results of validity and reliability test are shown in Table 1 were all indicators proved valid to be used as measuring instruments and all variables are also declared reliable. List of research variables, the value of validity test, and the mean of respondent answer are given in table 2.

Table 2. Research Result for Data testing and descriptive statistic

Indicator	Code	r	Cronbach Alpha	Mean	St. Dev
Consumer experience (PN)					
Consumer experience in buying eco-label products	PN1	0,773	0,884	3,490	0,977
Consumer Buying intention (KM)					
Consumers will buy eco-labeled products in stores	KM1	0,557	0,890	3,735	0,799
Consumers will try to buy fish products with eco-label fisheries	KM2	0,588		3,835	0,801
consumers will plan to shop so they can buy eco-label fishery products	KM3	0,437		3,685	0,836
Consumer Trust (KK)					
Consumers feel the role of BSN is now good	KK1	0,385	0,891	3,460	0,844
Consumers feel that BSN is an independent institution with no political interference	KK2	0,407		3,315	0,818
Consumers do not believe in BSN	KK3	0,391		3,330	0,695
Consumers have a positive perception towards BSN	KK4	0,648		3,535	0,782
Environmental campaign (KL)					
Environmental campaigns increase consumer knowledge of green products	KL1	0,699	0,888	3,945	0,797
Consumers enjoy the messages that environmental campaign ads pose	KL2	0,623		3,765	0,808
Environmental campaigns are able to guide consumers to make purchasing decisions on eco-labeled products	KL3	0,495		3,835	0,762
Variable Initiation of adoption process (MA)					
Consumers have heard of eco-labels on fishery products before	MA1	0,496	0,889	3,335	0,931
Consumers know the eco-label well	MA2	0,627		3,405	0,925
Process Variables Completing Adoption (SA)					
Consumers will strongly consider the eco-label when buying fishery products in general.	SA1	0,395	0,892	3,350	0,986

4.2 Descriptive statistics

Results of descriptive statistics show that all variables have a mean value above 3 and fall into the category agree. The highest mean value is environmental campaign variables with mean value 3,848. An environmental campaign is a campaign that aims to influence consumer purchase behavior by encouraging not to purchases products that pollute the environment and lead to buying products that environmentally friendly. The indicator that gets the largest mean value in the environmental campaign variable is KL1 which is 3.945. This means that respondents feel agree that environmental campaigns will be able to increase their knowledge about a green product.

4.3 Test the classical assumption

In this research there are two stages of the regression test, first is the initiation of an adoption process and second is the completion of an adoption process. The classical assumption test is performed on each of the multiple linear regression equations. All data used for factor-factor testing affecting the process of eco-label adoption of fishery products by consumers has passed the classical assumption test.

4.4 Hypothesis testing

Hypothesis testing is done to prove the hypothesis proposed in a study. In this study, there are 7 hypotheses. Hypothesis testing is done to determine the effect of independent variables on the dependent variable. Results of hypothesis testing give in table 3.

Table 3. The result of hypothesis testing

Dependent Variable	Independent Variable	β	Partial significantly	R	R square	Test Result
completion of the eco-label adoption process	initiation of an adoption process	1,027	0,000	0,839	0,704	accepted
	Consumer experience	-0,145	0,326	0,843	0,711	rejected
	Consumer Buying intention	-0,047	0,805	0,865	0,748	rejected
	Consumer Trust	0,43	0,087	0,846	0,716	accepted
	Environmental campaign	-0,325	0,095	0,843	0,711	rejected

Based on table 3, for the hypothesis test of initiation of an adoption process, the coefficient of determination is equal to 0.704 with the value of Sig. 0.000 or it can be said the completion of the process of consumer adoption of eco-label is significantly influenced by the initiation of the adoption process. Consumers buying intention rules as a moderator variable, based on the test results of consumer buying intention is a pure moderator variable with Sig value partially of 0.805. It cannot be used as an independent variable. Table 3 shows that variable consumer experience, consumer buying intention, and environmental campaign individually or partial have no significant effect on the completion of the eco-label adoption process.

Based on the results of testing the consumer experience is a pure moderator variable with Sig value. (Partial) of 0.326 so it cannot be used as an independent variable. The research finding that variables of consumer experience as variables moderator is in accordance with research conducted by Torgersen. Consumer experience is a pure moderator variable whose effect is moderate only and has no direct effect on the consumer for the completion of the eco-label adoption process.

Chase and Smith showed that 70 percent of respondents in a survey that claimed to be influenced by environmental campaigns in the decision to buy products to eco-label [12]. In addition, research conducted in Malaysia suggests that research on green product purchases should add environmental campaign variables as they may affect consumers' buying behavior and willingness to pay more for eco-labeled products. Environmental campaigns allow consumers to recognize the benefits of purchasing eco-labeled products and will significantly improve the green market [13].

5 Conclusion

Completion of the process of consumer adoption of eco-label is significantly influenced by the initiation of the adoption process. Consumer experience, consumer buying intention, and environmental campaign individually or partial have no significant effect on the completion of the eco-label adoption process. Acceleration the adoption process that is consumer movement from initiation phase to completion phase has moderator variable that gives most significant influence. The moderator variables are consumer buying intention with influence value that is equal to 0,748 and consumer experience with influence value that is equal to 0,711.

References

1. UNEP. Overfishing, a major threat to the global marine ecology, Environment Alert Bulletin, UNEP United Nations Environment Program (2001)
2. WWF. Profile of Fisheries and Marine Indonesia, Entering Sustainable Management Strategy, *Information sheet* January edition (2008)
3. GEN. Introduction to Ecolabelling. GEN: Global Ecolabelling network Information paper (2004)
4. C.R.Wessells, K.Cochrane, C.Deere, P.Wallis, R.Willmann, FAO Fisheries Technical Paper 422 (2001)
5. D. Brechard, *J Ecological Economics* Vol **69** 115–125 (2009)
6. W.Goyert, R. Sagarin, J. Annala. *J Mar. Policy* Vol. **34**, 1103–1109 (2010)
7. S. Mc Donald, C.J. Oates, P.J. Alevizou, C.W. Young, K. Hwang, *J Marketing Management* Vol. **28** 445-468 (2012)
8. J. Thogersen, P. Haugaard, A. Olesen, *European Journal of Marketing* 44 (11/12), 1787-1810 (2010)
9. B. Chekima, S. A. W. S. Khalid Wafa, O. A. Igau, and S. Chekima. *J Asian Soc. Sci* Vol. **11**, no. 10, pp. 318–329 (2015)
10. K.G. Grunert and J.M. Wills. *J. Public Health (Bangkok)* Vol **15** 5 385–399 (2007)
11. R. Purwaningsih, A. Susanty, A.K. Wafa, *Proceeding* of National Conference of Indonesia Industrial Engineering higher education provider cooperation, Malang, East Java (2017)
12. Chase, D., & Smith, T. K. *J Advertising Age* Vol **63** (1992)
13. Gutierrez, A. & Thornton, T. F. *J Sustainability* Vol. **6** 8195-8217 (2014)

Analysis of Seafloor Sediment Distribution using Multibeam Backscatter Data

Danar G Pratomo[1,*], *Khomsin*[1], *Mokhammad N* Cahyadi[1], *Kamila* Akbar[1], and *Evasari* Aprilia[1]

[1]Geomatics Engineering Department, Faculty of Civil Engineering, Environment, and Geo-Engineering, Institut Teknologi Sepuluh Nopember, Surabaya, Indonesia

Abstract. Multibeam sonars use an acoustic wave to provide a view of seafloor topography and collect its backscatter. Seafloor backscatter image can be used as a means of quantitative classification of seafloor lithology. This allows sediment expert to examine spatial distributions of seafloor sediment types and discriminate among them. This method is expected to reduce expensive bottom-sampling programs. The research examined multibeam data collected from Indonesia Exploration 2010 Project. This project employed a Kongsberg EM302 multibeam system to collect seafloor topography and backscatter data at Sangihe Talaud Sea, North Sulawesi. The average depth of the area study is approximately 2400 m. The research applied the Angular Response Curve (ARC) method to analyze the echo strength of the backscatter as a function of the grazing angle. The range of backscatter intensity in this area is from -9dB to -67dB. Based on the backscatter data, there are five types of sediment in this area: clay, mud, gravel, sand, and boulder.

1 Introduction

The multibeam echo-sounder (MBES) systems use an acoustics signal to reveal unprecedented detail of the bed morphology and bed form migration, whilst the backscatter signal from the bed can also be utilized to provide information on the bed roughness [3]. MBES systems also provide acoustic returns from within the water column that can image within the flow, and this ability has recently been applied for both object detection and visualization [4]. The acoustic wave emitted by a MBES reflects from obstacles of various natures in the water column and at the boundary interfaces of the medium. The returning energy from the seafloor, called backscatter, is related to the physical nature of the sea bottom as well as to some characteristics of the transmission and reflection (e.g., the angle of incidence and the pulse-length) of the incident signal [2].

The backscatter intensity of the acoustic signal is reflected by the seafloor follows the angular response function which shows the relationship between the value of the backscatter intensity and the angular response. This relationship can be used to determine the seafloor sediment type on the seafloor and to identify the type and distributions of the

* Corresponding author: guruh@geodesy.its.ac.id

morphological segregation of the seabed sediments [1]. The backscatter imagery provides a representation of the seafloor which is a mixture of both bathymetric and sedimentological signatures. The acoustic response of seabed varies as the sediment type, roughness, grain-size, compaction, and slope change. In addition, returned backscatter energy depends not only on the seafloor physical properties but also on the sonar configuration, water column propagation, and measurement geometry.

The geometric and radiometric corrections are necessary to get backscatter data which represent an inherent property of seabed that reflects its physical properties. In this case, to a first order, the geometric and radiometric corrections have been applied by the manufacturer [8]. A number of different approaches to characterization via backscatter have been developed with the primary aim of acoustic segmentation; that is, mapping discrete geographical areas of the seafloor with the same acoustic signature. These techniques have also been utilized to extract different properties from the backscatter data [5].

The research focused on backscatter data analysis to identify sediment distribution in Sangihe Talaud water, North Sulawesi. The research analyzed a relationship between the backscatter intensity value and the angular response. The backscatter mosaic results generated from the acoustic signal processing are inversely proportional to the given spatial resolution. When an angular curve produces a high value, the resulting spatial resolution will be low [3]. The research examined multibeam data which is collected from Indonesia Exploration 2010 Project. This project utilized a Kongsberg EM302 multibeam system to acquire seafloor topography and backscatter data.

2 Data and Research Area

The research used multibeam data from the Leg 2 of the Expedition Indonesia Exploration 2010 Project. This project used an EM302 Kongsberg system which is attached at vessel survey RV Okeanos Explorer. This 30kHz multibeam system is designed to perform seabed mapping with high resolution and accuracy to a maximum depth of more than 7000m. The system has up to 432 soundings per swath with pointing angles automatically adjusted according to achievable coverage or operator defined limits. The multibeam system equipped with dual swath (two swaths per ping) system. The transmit fan is duplicated and transmitted with a small difference in along-track tilt. The applied tilt takes into account depth, coverage and vessel speed to give a constant sounding separation along-track. In dual swath mode, 2 swaths are generated per ping cycle, with up to 864 soundings. The beam spacing is equidistant or equiangular.

EM302 uses both CW and FM pulse forms. FM sweep with pulse compression on reception is used to increase the maximum useful swath width. The FM pulses will also increase the resolution compared to CW pulses of the same length. The transmit fan is split in several individual sectors with independent active steering. This allows stabilization which compensates for the vessel movements: yaw, pitch and roll. Each transmit sector has individual beam focusing. The EM 302 transducers are modular linear arrays in a Mills cross configuration with separate units for transmit and receive [10].

The EM302 was utilized to collect the data from June 23rd to June 25th, 2010. A sea level observation was also conducted to correct the depth measurement from multibeam. This observation is located at Port Bitung station and started from June to July, 2010. A Sound Velocity Profiler measurement was also conducted on vehicle during MBES acquisition. The research area is located in west side of Sangihe Island, North Sulawesi Province. The area is bound with geographical coordinates 2°01' LU - 6° 24' LU and 124° 45' BT - 128° BT. The average depth of the area study area is approximately 2400m. Fig. 1

shows the location of the research area. The research utilized Swathed, a MBES processing software developed by Hughes Clarke from UNH.

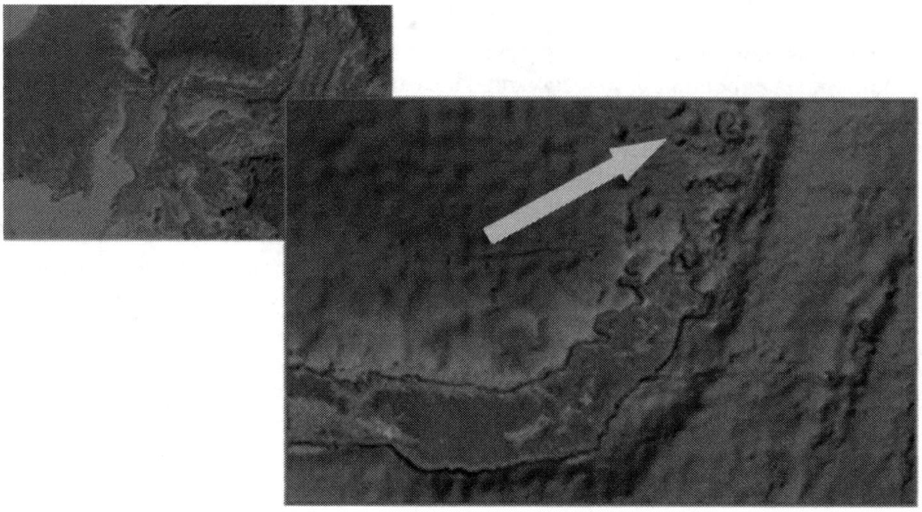

Fig. 1. Map of the research area. Blue arrow shows the location of study site in Sangihe Talaud Sea, North Sulawesi.

3 Methodology

Bathymetric data collected using the Kongsberg SIS software is fully corrected for ship's motion, navigation and sound velocity [9]. Application of corrected tides and bathymetric data cleaning are processed using the Swathed and tide reference data provided by BIG (Indonesia Geospatial Information Bureau). Next, is creating a gridded bathymetry surface using processed data. In this research, backscatter data were processed using the Swathed as well. The backscatter data is imported directly from the datagrams. The data used in this research are from line 15 to line 31. Data acquisition was performed two days from 23-24 June 2010 by R/V Okeanos Explorer. Raw data is then processed through unravel process and convert to Ocean Mapping Group format: *.merged is the format for bathymetry data and *.ss is for backscatter data.

Tidal correction is required to achieve the depth even though it is not essential as the area is classified as deep water. The tidal data during the acquisition of the MBES data is adjusted to *.bin format to be corrected with the *.merged file. The MBES data cleaning process is performed on the Swathed profile tab with the median filter which can eliminate spikes with a significant different depth value from the median depth value. After the data free from spikes, a Digital Terrain Model (DTM) was created using parameters as follow: 5m per pixel, beam width 1.0°, vertical exaggeration 1.0. The DTM file is in r4 format then is exported to ESRI format with R4toESRI command to be processed further. The DTM created from this data can be seen in Fig 2.

In order to achieve scattering values, the backscatter is corrected to remove nadir stripping, eliminating beam pattern variation, and georefencing. The next step is creating a backscatter mosaic with mos2 command. The mos file is then exported to ESRI format with the r4toESRI command. The backscatter mosaic file is processed with ArcMap to classify the sediment type of the backscatter value obtained from the pixel value (DN = Digital Number) in the mosaic which is then converted to 24-bit decibels (dB). The decibel and transmit angle values of MBES (grazing angle) are then made graphs to obtain the backscatter strength and the sediment type relationships.

The following are the corrections steps foresaid to achieve a corrected backscatter image: [a] correction of transmission losses associated with absorption and oblique range, [b] application of calibration gains, and [c] correction for the instantaneous ensonified area. The output of this reduction process is considered as the best estimate of the absolute backscatter angular response for the survey area.

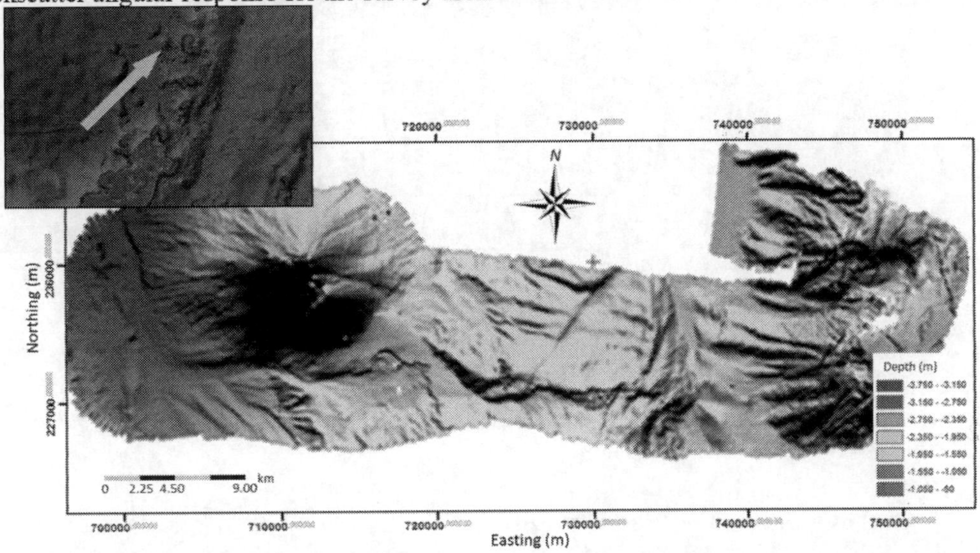

Fig. 2. The Digital Terrain Model (DTM) of the research area created from MBES data. The location of the DTM is pointed by blue arrow.

4 Results and Analysis

4.1 Bathymetry and tide

Horizontal positioning method used in the research is a differential positioning using a C-NAV DGPS system. This system can provide a 2.0m accuracy. Most of datasets used in this research were collected by Kongsberg EM302 multibeam system. In the processing of MBES raw data, cleaning data used to eliminate the noise and spike. This can be from several factors such as under water biota activity, water bubbles, and reflective objects the MBES acoustic signal before the signal arrives at the bottom of the waters.

In order to get the depth, the MBES data then corrected with the tides data. The tide used in the research is observed tide from Bitung Station, North Sulawesi. The duration for the tide observation is 1 month during the time of MBES data acquisition (June 24th - July 24th, 2010). The tide observation can be seen in Fig. 3.

 https://doi.org/10.1051/matecconf/201817701026

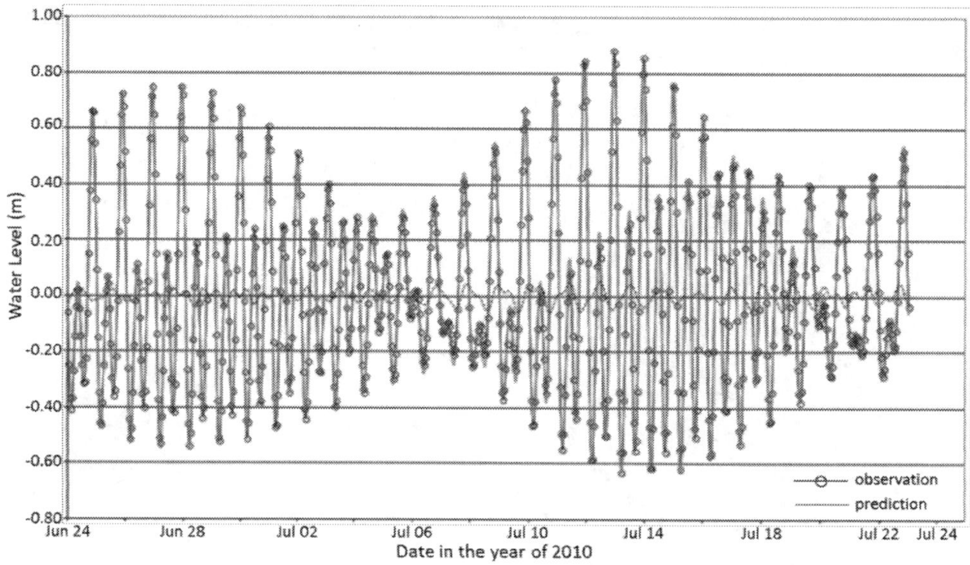

Fig. 3. Tide observation in the research area. The observed tide is represented by blue line, and the magenta line represents the tide prediction. The black line is the difference between the tide observation and prediction.

4.2 Analysis of backscatter strength

After the correction step for each raw data line, then each line is combined to develop a backscatter mosaic. The classification of the sediment types was conducted by looking at the backscatter strength of the echo return in the mosaic. The strong scatter is represented by black color and it shows hard sediments. Vice versa, the weak scatter is represented by white color and it shows soft sediments. Based on the backscatter mosaic, the range of backscatter strength is from -10dB to -67dB. Furthermore, this mosaic is processed using Arc map to classify the sediment on the seafloor. The backscatter strength mosaic can be seen in Fig. 4.

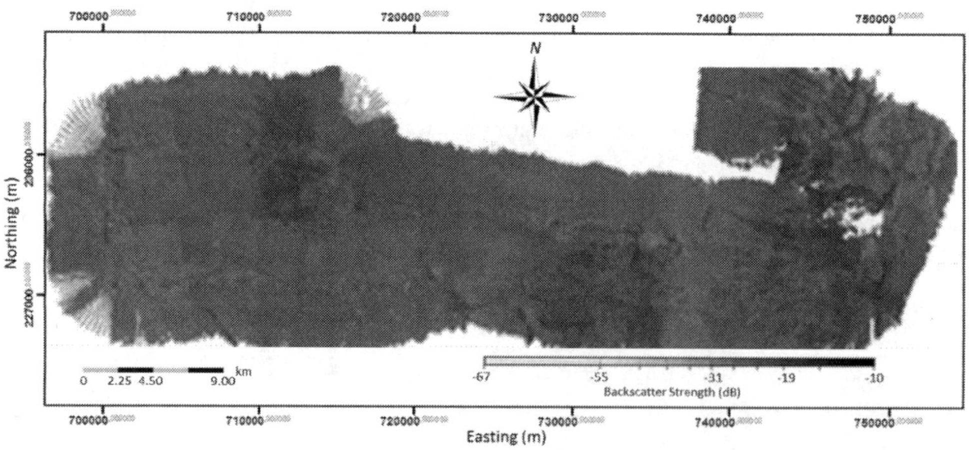

Fig. 4. The backscatter image of the research area. The backscatter strength is represented by grey scale color. White color represents a weak intensity and black color represents a strong intensity.

5 Discussions

5.1 Sediment distribution based on backscatter value

The backscatter strength is in a Digital Number. The magnitude of the backscatter strength is derived from three beam sweeping zone of the MBES which are the nadir zone, the oblique corner zone, and the angular critical zone. In the area which is composed by rock outcrop has a very strong scattering value (high intensity). This intensity ranges from 237DN to 211DN or from -9dB to -67dB in decibel unit. The area composed by boulder has a backscatter value from 237DN to 217DN or from -9dB to -18dB. The areas composed by gravel have a backscatter value from 209DN to 187DN or -23dB to -34dB. The area which is dominated by sand has a backscatter value from 185DN to 159DN or -35dB to -48dB in decibel unit. The area contains mud has backscatter value ranges from 157DN to 137DN or from -49dB to -59dB. Area composed by clay has very low backscatter strength (low intensity) from 135DN to 127DN or -60dB to -67dB.

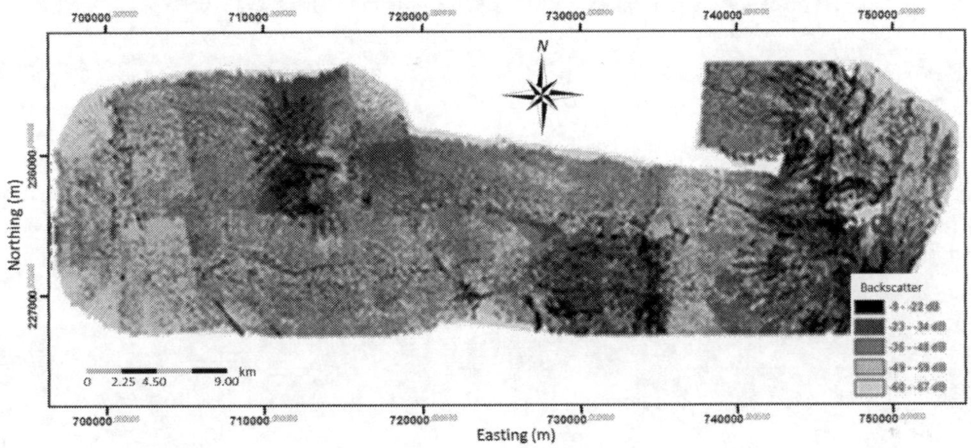

Fig. 5. Distribution of backscatter value. The backscatter values is classified into 5 types.

Fig. 5 shows the classification of the backscatter values. The values range from -9dB to -67dB, these values have been converted from DN value. The range for the backscatter value is classified into 5 classes according to the type of sediment composed the seafloor. The black color in the mosaic shows the value a strong backscatter which indicates a hard sediment type. While the bright color in the mosaic indicates the backscatter value is a soft sediment type.

Fig. 6 shows the backscatter strength with a range of values -9dB to -67dB. There are 5 types of sediment which covers the seafloor. Based on the sediment classification, the research area is dominated by mud with backscatter values ranges from -49dB to -59dB.

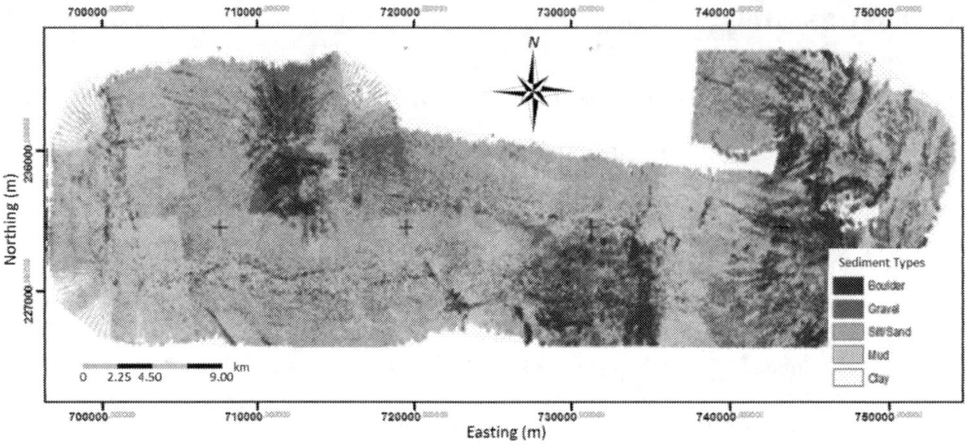

Fig. 6. Sediment classification based on backscatter strength. There are five types of sediment cover the seafloor.

5.2 Angular Response Curve

The corrected correlation obtained from the extraction of backscatter values of each sampling line samples are then plotted in the reflection angle / grazing angle curve in order to determine the relationship of the sediment type based on the MBES reflection angle to the bottom surface of the waters.

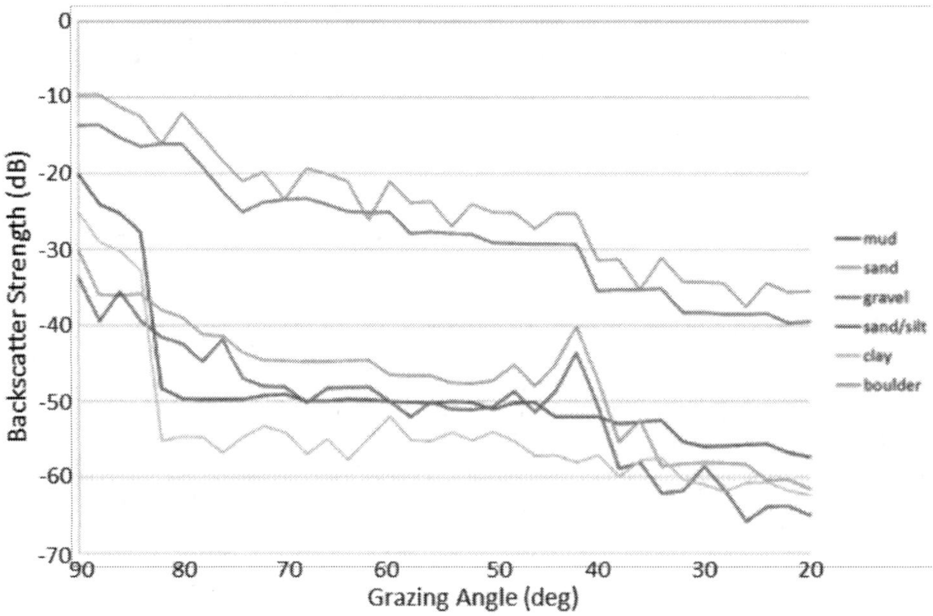

Fig. 7. The relation between the grazing angle and the backscatter strength. Generally, as the grazing angle decreases, the backscatter intensity will also decrease. The maximum backscatter strength is in nadir.

Fig. 7 shows the relationship between the backscatter strength and the grazing angle. The different color on the curve represents the backscatter strength of each sediment types.

The response of the backscatter strength to the grazing angle on the nadir is very strong due to the surface flattening of the gravel sediments which allows reflecting the acoustic signal by scattering the acoustic signal. However, when grazing angle is 70°-60° there is a significant decrease in backscatter value. This occurs due to the oblique angle of the acoustic signals received by the transducer. At the critical angle is from 30° to 10°, the acoustic signal is reflected in a small angle so that the backsliding value is weakened or even lost.

Similar to the angular response curve of the gravel sediment type, the angular response curve of the sand sediment type has a very strong backscatter value at nadir, and decreases in the slope angle area or 70°-40° grazing angle. However, at the critical angle, the backscatter value increases and forms a peak curve at about 30° of grazing angle. For the mud sediment type, at the nadir angle, the backscatter value still has a strong intensity dominated by scattering volume, but when the grazing angle is 80° and declining, the mud-backs backscatter value weakens with a smoothly decreasing curve at 40° grazing angle.

6 Conclusions

The analysis of the relationship between backscatter strength and grazing angle is confirmed as a promising technique for a quantitative automated processing of large datasets of seafloor backscatter. This research successfully applied this technique. The availability of absolute calibrated process with appropriate data reduction was the key to the success of this approach. The result from the research shows a good clustering of different sediment types. This could be a relevant basis for a reflectivity angular response types to classify the sediment distribution. Distribution of corrected scattering value in the research area has a maximum repository value of -10dB and a minimum of -67dB and it has a DN range from 237DN to 121DN. The results of the distribution of sediment types based on the analysis of backscatter value using an Angular Response Curve method is as the following: [a] -9dB to -22dB is boulder, [b] -23dB to -34dB is gravel, [c] -35dB to -48dB is sand, [d] -49dB to -59dB is mud, and [e] -60dB to -67dB is clay.

The authors would like to thank BPPT and NOAA as the data provider and Dr. Hughes Clarke for the Swathed, a multibeam processing software.

References

1. R. Che Hasan, D. Ierodiaconou , L. Laurenson, A. Schimel. Integrating multibeam backscatter angular response, mosaic and bathymetry data for benthic habitat mapping. PLoS One. 9(5): e97339. 10.1371/journal.pone.0097339 (2014)
2. C. De Moustier. Angular dependence of 12 kHz sea-floor acoustic backscatter. Journal of Acoustical Society of America, pp. 522-531 (1991)
3. L. Fonseca, L. Mayer. Remote estimation of surficial seafloor properties through the application angular range analysis to multibeam sonar data. Marine Geophysical Researches. 28(2): 119-126. 10.1007/s11001-007-9019-4(2007)
4. M. Giuseppe, B. Calder B. Remote identification of a shipwreck site from MBES backscatter. Journal of Environmental Management Volume 111, 30 November 2012, Pages 44-52 (2012)

5. E. Hammerstad. Backscattering and seabed image reflectivity, EM Technical Note (2000)
6. J. E. Hughes Clarke. Towards remote seafloor classification using the angular response of acoustic backscatter: a case study from multiple overlapping GLORIA data. IEEE Journal of Oceanic Engineering, Volume 19, pp. 112-127 (1994)
7. H. Manik, D. Yulius. Development and application of MB system software for bathymetry and seabed computation. International Journal of Software Engineering and Its Applications (2015)
8. D. G. Pratomo. Coupling of repetitive multibeam surveys and hydrodynamics modelling to understand bedform migration and delta evolution. The University of New Brunswick: Canada (2016)
9. SIS. Seafloor Information System, Kongsberg ME70 Operator Manual. Kongsberg Maritime (2010)
10. X. Lurton. An Introduction to Underwater Acoustics: Principles and Applications (second ed.), Springer, published in association with Praxis Publishing, Chichester, UK, Heidelberg; New York (2010)

Structural Analysis on the Block Lifting in Shipbuilding Construction Process

Muhammad Nurul Misbah[1,*] , *Septia* Hardy Sujiatanti[1] , *Dony* Setyawan[1], *Rizky* Chandra Ariesta[1], and *Satriyo* Rahmadianto[2]

[1]Department of Naval Architecture, Institut Teknologi Sepuluh Nopember, Surabaya, Indonesia
[2]PT Dok dan Perkapalan Surabaya (Persero), Surabaya, Indonesia

Abstract. Deformation is unavoidable in some stages of the ship production process. Deformation occurs when the blocks are lifted in an erroneous eye pad position. The aim of this study was to determine behaviour of the ship structures during assembly and lifting process. A typical tanker was used as the basic structural shape. The bottom structures was modeled to be analyzed. In this study, deformations and stresses on the bottom structures during the block lifting are investigated using the finite element method. The deformations and stresses are evaluated and critical condition detected. The preferable block lifting method with the minimum distortion on the bottom structures of tanker is proposed.

1 Introduction

Ship, as a complex product, was divided correspond to different criteria. One of the important criteria is spatial criterion. This criterion is important to the production phase and it is dependent on the technological possibilities of the shipyard. Basic of spatial division of the ship, widely used for most types of ships, is division on macro spaces i.e. after peak, machinery space, cargo space, fore peak, and superstructure. Division in macro space also well-known as a block or grand block system in shipbuilding process.

Block assembly is the process in which the hull of a ship is built by joining blocks together. It is also called hull assembly. Blocks are large welded steel structures that are typically manufactured as single part. When a block is ready, it is typically arranged near the launching place, lifted to the launching place, and welded to other blocks there. Block assembly in shipbuilding is a unique phase when shipbuilding is compared with the normal type of mechanical engineering production [1].

Currently, many shipyards use the method of block assembly in the shipbuilding process. This method makes the ship to be produced faster than the conventional method of shipbuilding. In the method of block assembly, ship production process is

* Corresponding author : mnmisbah5@gmail.com

done by combining the blocks into a large block or part of the ship [2]. The problem occurs when the process of block assembly is after the lifting of the block. The lifted block will changes shape or deformation [3].

In this paper, the bottom structures of typical tanker is carried out to be analyzed. The typical bottom structures of tanker are given in Fig.1. The analysis of block lifting on bottom structures of tanker using the finite element method. The lifting process of the tanker bottom structure block was simulated by install some eye plates or eye pads on portside block and center block with various position of eye pads. The goal of the finite element analysis was to successfully predict the effect of block lifting in shipbuilding process with various position of eye pads. The optimum of eye pads position can be assumed based on the minimum deformation resulted from lifting process.

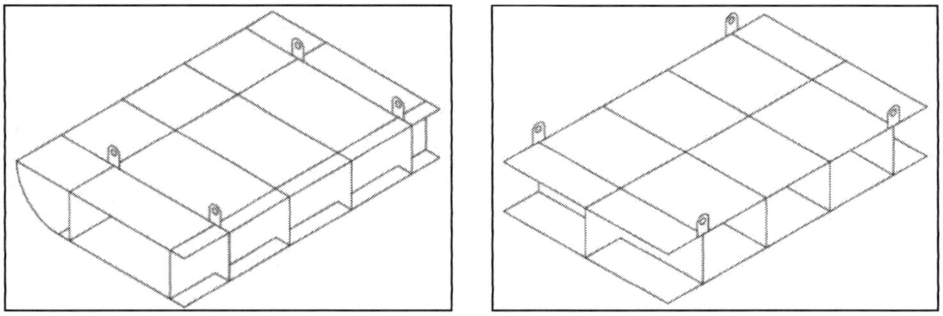

(a) (b)

Fig. 1. Bottom structures of tanker at portside block (a) and center block (b)

2 Shipbuilding Production Process

Ship production is a very difficult job as it covers many processes. A ship is produced by performing thousands of work activities. To produce ships, various types of workstations are required. Each workstation has a duty to ship production.

Assembly is one of the most common operations in the production process. This involves combining parts in a separate and inseparable way. As a result, a complicated product called assembly is made. The singularity of the ship's structure is connected with a large number of elements [4]. The hull production consists of combining with the welding method in an inseparable way. The hull, as the final assembly, is a unified whole. This is a distinctive feature that distinguishes the hull assembly from mechanical assemblies. In the case of the shipbuilding process, the component may be called a sub-assembly, i.e. a previously assembled structure.

Components can be assembled in different sequences called sequences. Each assembly sequence leads to the same result - the final structure. However, the sequence differs in terms of the value of many criteria, for example: implementation difficulties, energy costs, employee safety, qualifications required for welders, total execution time, etc. The selection of the best sequence is therefore a multi-purpose optimization problem. The whole class of problem types and their settlement methods are called assembly sequence planning (ASP).

3 Finite Element Model

A typical bottom structure block is considered. The block size and topology is defined by shipbuilding assembly practice and drawings of the ship, designed and produced in a shipyard. The bottom portside block dimensions are 30.5x12x6 meters (width x length x height). Block consists of bottom and inner bottom structures. Bottom structure consists of bottom plate, bottom longitudinal, and bilge plate. Inner bottom structure consists of inner bottom plate and inner bottom longitudinal. Plate floor, bottom center girder and bottom side girder are installed between bottom plate and inner bottom plate.

The finite element analysis is carried out to analyze the stress and maximum deformation during block lifting process. The bottom structure of tanker was modeled using finite element software package. In the analyses, three-dimensional finite element models were developed. Eight node shell element having six degrees of freedom per node was used to model the steel plates and three node beam element with six degrees of freedom per node was used to model the stiffeners [5]. The bottom structure models were analyzed for various eye pad position. The variations of eye pad positions for portside block are shown in Fig. 2.

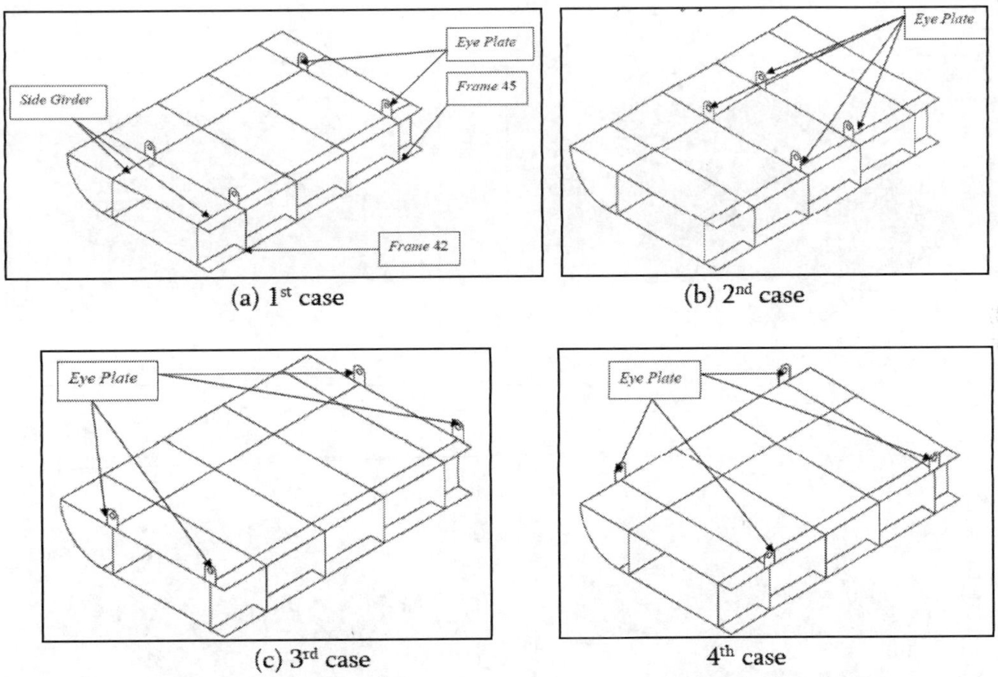

(a) 1st case (b) 2nd case

(c) 3rd case 4th case

Fig. 2. Eye pad position at the portside block

The variations of eye pad positions for center block are shown in Fig. 3.

 https://doi.org/10.1051/matecconf/201817701027

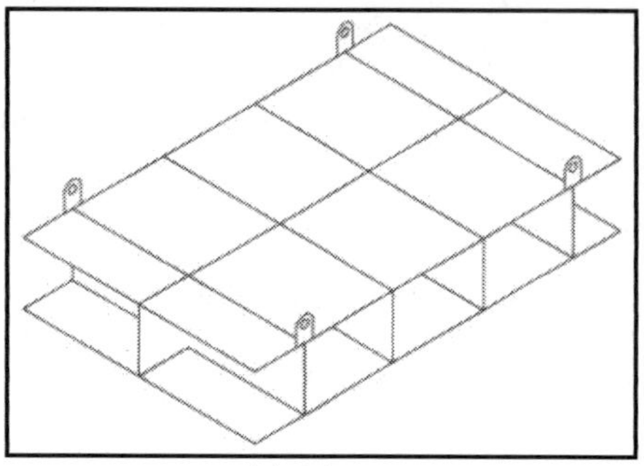

(a) 1st case

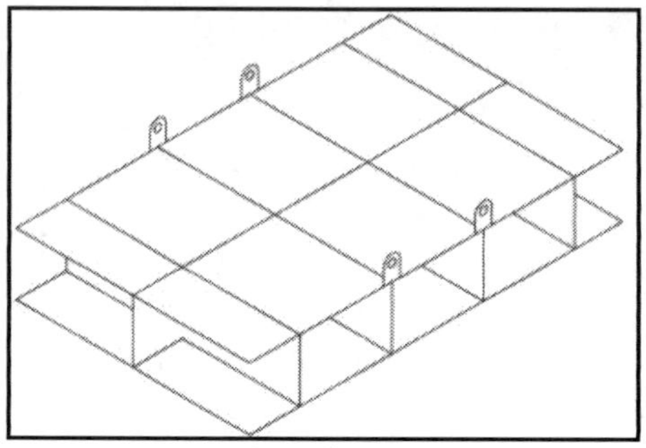

(b) 2nd case

Fig. 3. Eye pad position at the center block

4 Analysis of the Block Lifting

By lifting the block, all the weight will be transferred to the crane through, in this case, four eye pads or lifting pads. The boundary conditions fix all the lifting pads. Caused by that, the weight will cause displacement mostly in the vertical direction, i.e. the y-axis. Fig. 4 presents deformation of the bottom structure when the portside block is being lifted by various eye pads positions.

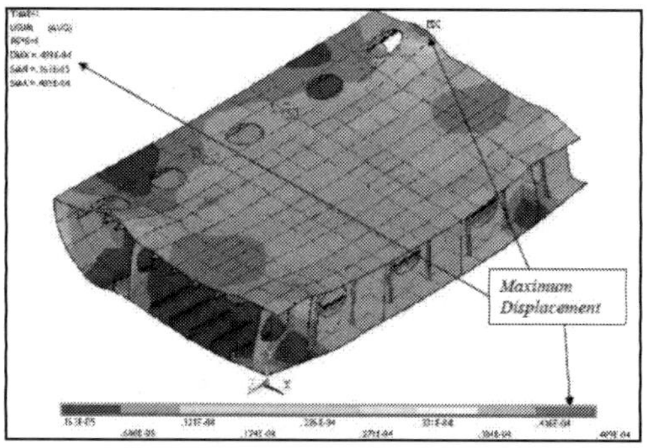

1st case

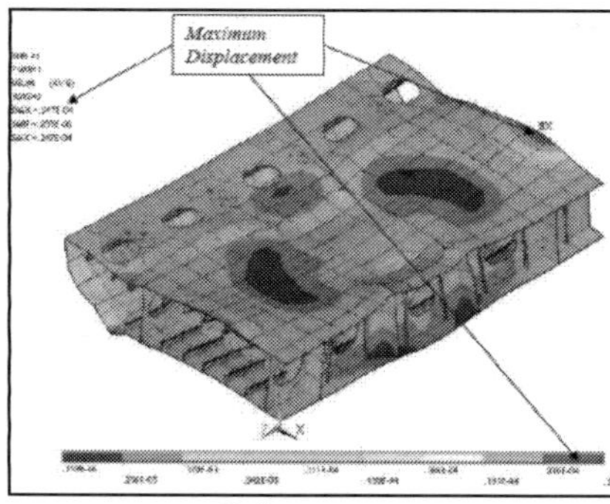

2nd case

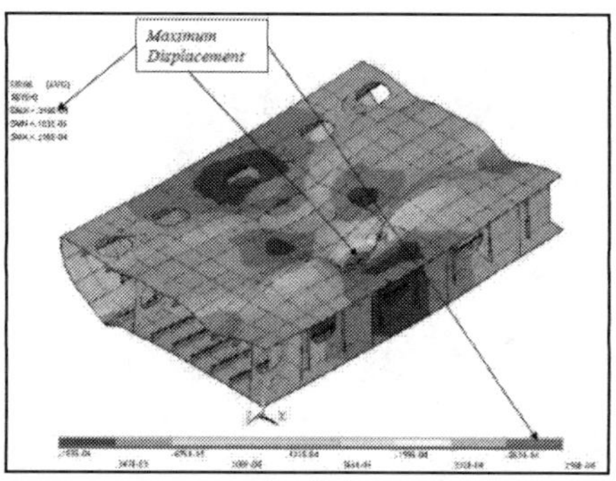

3rd case

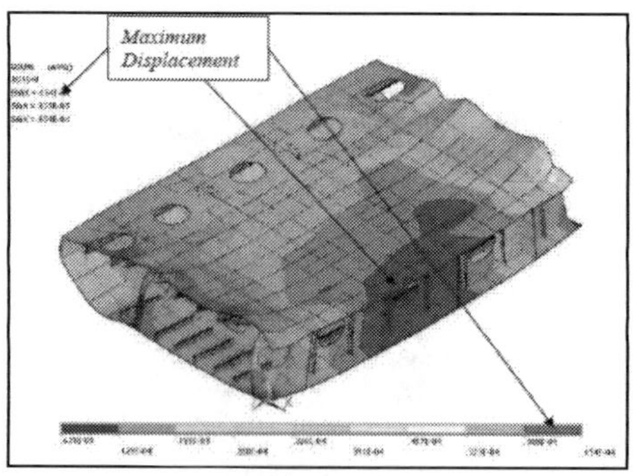

4th case

Fig. 4. Deformation at the portside block

The deformation of the bottom structure when the center block is being lifted by various eye pads positions are shown in Fig. 5.

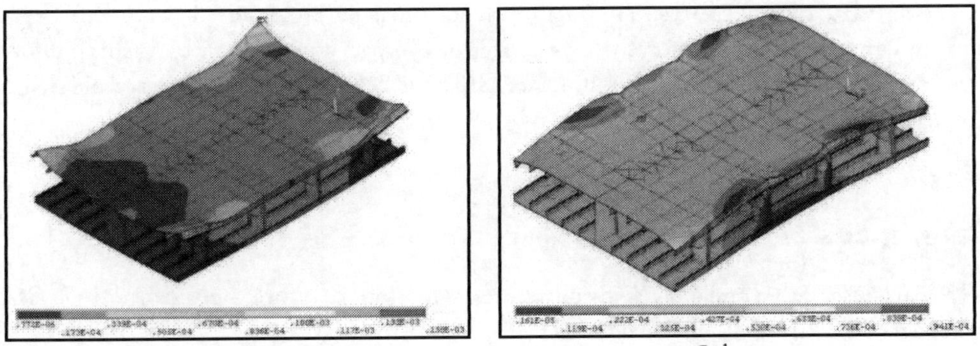

1st case

2nd case

Fig. 5. Deformation at the center block

5 Result

The simulation results of bottom structures with various eye pad positions on portside block show the maximum deformation. The finite element simulation results have shown in Fig. 4. and Fig. 5. The maximum deformation on the portside block of bottom structure for various eye pad positions are given in Table 1.

Table 1. Maximum deformation for various eye pad at the portside block

Type	Max. Deformation [mm]
1st case	0.049
2nd case	0.025
3rd case	0.030
4th case	0.065

The maximum deformation on the portside block of bottom structure for various eye pad positions are given in Table 2.

Table 2. Maximum deformation for various eye pad at the center block

Type	Max. Deformation [mm]
1st case	0.0412
2nd case	0.0928

6 Summary

According to the analysis results presented in Fig. 4 and Fig. 5 also presented in Table 1 and Table 2, it can be concluded that the lowest deformation occur on the block of bottom structure is 2nd case i.e. 0.025 mm for portside block and 1st case i.e. 0.0412 mm for center block. Therefore, the variation of eye pad position can be recommended for block lifting process on bottom block of tanker is 2nd variation for portside block and 1st variation for center block.

References

[1] Varghese, R., Yoon, D. Y, Shipbuilding erection network optimization: A TSP method, Journal of Ship Production, 22, 3, 139–146, (2006)

[2] R. R. Iwańkowicz, Integration Of Ship Hull Assembly Sequence Planning, Scheduling And Budgeting, Advances in Science and Technology Research Journal, Volume 9 No. 25 pp 12-19, (2015)

[3] V Adan, C Alexandra, B Juan, S Hishashi, Numerical Analysis of Straightening Process of Thin Plate Structures by Elastic FEM Based on the Inherent Straint Method, Transaction of JWRI vol.41 (2012)

[4] M Ozkok, Risk Assessment In Ship Hull Structure Production Using FMEA, Journal of Marine Science and Technology, DOI: 10.6119/JMST-013-0222-1

[5] S H Sujiatanti, A Zubaydi, A Budipriyanto, Finite Element Analysis of Ship Deck Sandwich Panel, The 1st SENTA International Conference on Marine Technology, ITS Surabaya (2016)

Cost and time analysis of graving dock project

Silvianita[1,*] , *Nur* Aprillia[2], *Yeyes* Mulyadi[1], *Wahyudi Citrosiswoyo*[1], Suntoyo[1]

[1]Ocean Engineering Department, Faculty of Marine Technology, Institut Teknologi Sepuluh Nopember, Surabaya, Indonesia
[2]Graduate Student, Ocean Engineering Department, Faculty of Marine Technology, Institut Teknologi Sepuluh Nopember, Surabaya, Indonesia

Abstract. The construction of new shipyard in the Cilacap area is one of the best ways to reduce the accumulation of vessels. The construction of shipyard has many risks, one of the risks that can occur during the construction of the shipyard is in terms of cost and time. Project scheduling is a tool to make sure the project finished on time. This paper discusses project scheduling and cost calculation using Critical Path Method (CPM) and Program Evaluation Review Technique (PERT) method for Graving Dock project development in Cilacap. Project scheduling is useful to indicate the relation of each activity to another. This paper will discuss the critical activities during Gravity Dock development and the duration for the whole project using CPM. The next step is to determine the probability of the completing project that can be finished within the time frame using PERT. Based on the CPM analysis, the probability of Graving Dock project can be finished in 204 days is 79,81 %. The total cost for Graving Dock project in Cilacap is Rp. 6.486.122.769.

1 Introduction

Indonesia is the largest archipelagic country in the world which has the largest maritime area three times than its land area or close to 70% of the total country of Indonesia. Therefore the marine territory of Indonesia can be utilized as much as possible with a lot of potential to have a maritime industry and a large and powerful shipyard.

A shipyard is a place used for making and repairing a ship [1]. Great shipbuilding production in shipyard can be achieved by good internal communication and coordination and also understand of modern shipbuilding requirements and characters by shipyard company. Before undertaking a production scheduling process, firstly it would be better to carry out a survey to investigate the situation and condition of the shipyard namely human resources, information systems, shipbuilding technology, raw material procurement and shipyard operation [2].

The project is an activity with a certain allocation of resources and purpose to carry out tasks with limited time whose targets have been clearly outlined [3]. Project are generally limited by time and cost [2]. The success or failure of a project is often caused due to fail of coordination so that the activities become not efficient [4]. Those matters could lead the delay of the project completion. Project scheduling is very important to make sure the

* Corresponding Author : silvianita@oe.its.ac.id

project can run smoothly [5] and the application of project scheduling in various shipyard has been carried out [6-8]. Currently there are many new shipyards have been constructed in order to distinguish themselves from and to produce system innovations [9]. The project required a network planning to match the scheduling of time and cost that have been determined in order to make sure the project is on schedule and there is no additional time and cost [10].

Network planning will be useful to manage the project within the time frame. If there is any delay, it will affect the duration of the project and the budget that is not in accordance with the contract [11]. The shipyard development project aims to repair or maintain tug boat, vessel owned by government, private company etc operating in Cilacap area. Since the shipyard in Cilacap still very few therefore it is needed for consideration of the new development of shipyard to do repair and maintenance of the vessel.

2 Methodologies

The methodologies used in this paper are generated based on CPM and PERT method. The objectives of this paper namely determine the duration, cost and also the probability of developing Graving Dock can be finished on time.

2.1 Data Collection

Data collection used in this paper is based on shipbuilding project namely:

- Data duration schedule of project work
- Data job volume

2.2 Project Scheduling Using CPM Method

Project Scheduling using the CPM method is a network analysis facility that seeks to optimize project cost through reduction or acceleration of project completion time. This stage begins by making the network of work breakdown structure based on the data obtained. Then determine the critical path by performing forward calculations, backward calculation and calculating the total float so that the critical path can be identified and the total duration obtained.

2.3 Probability of Project Finish on Time

In order to identify the probability of project can be finished on time then PERT is used. PERT is useful to plan the project and to minimize the delay, as well as disruption of production and coordinate various parts of a job as a whole and accelerate the completion of the project. The PERT technique generates a controlled and orderly job because the schedule and budget of a job are predetermined before the project takes place.

2.4 Project Cost Analysis

After conducting analysis by using CPM method, then it can be seen the total duration of the project from the duration of the critical path. Cost is one of the things that must be prepared because it is needed as a guide in project implementation.

3 Results and Discussion

3.1 Project Planning

Construction project development Graving Dock in Tanjung Intan, Cilacap is a project organized by one of the shipyard company in Cilacap. This project takes 204 days work time from 12 January 2018 until 31 July 2018.

The project starts by planning the time and cost required for graving dock, and continues with the activity that needs to be done during the time frame. The estimation duration of each activity in graving dock can be based on the previous project experience.

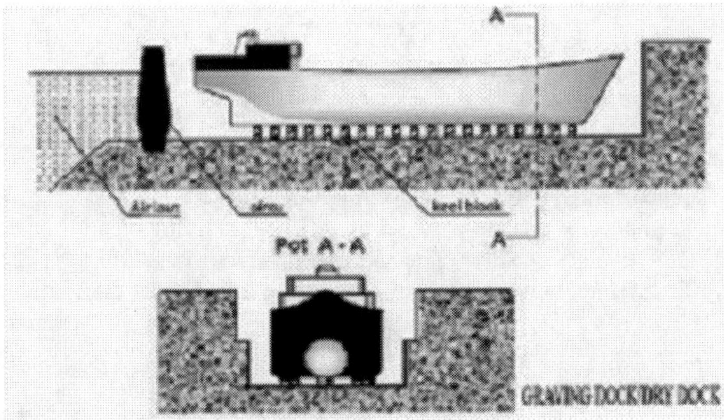

Fig 1. Graving Dock

3.2 Scheduling Use *Critical Path Method*

This stage begins with network planning to schedule the activity using CPM (Critical Path Method) based on WBS data. Figure 2 shows the network planning for this project:

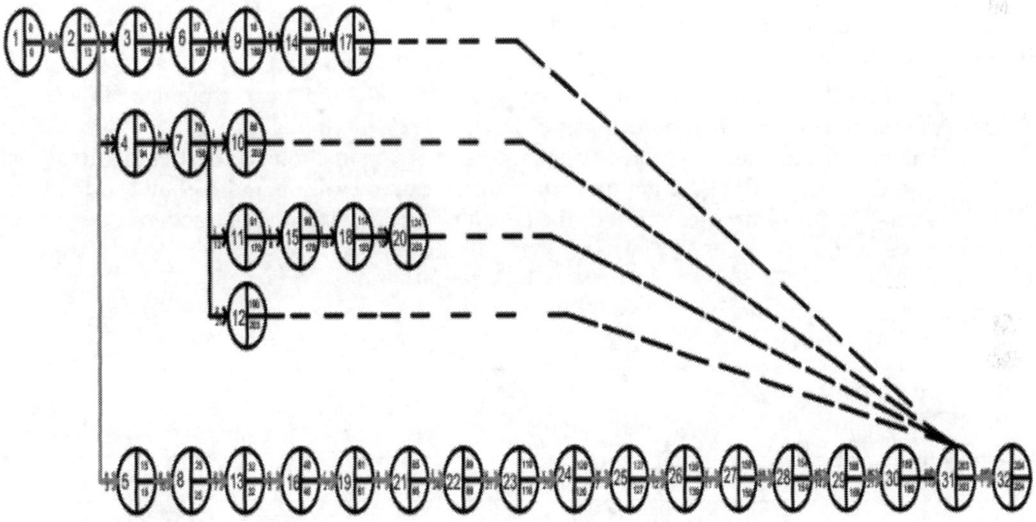

Fig 2 .*Network Planning* CPM

Based on network planning CPM, it was found that the length of duration required to complete the project was 204 days. Based on the Fig 2. the red line is a critical activity.

3.3 Scheduling use *Program Evaluation Review Technique* Method

Scheduling using PERT method is divided into several steps as follows [5]:

- 3 estimation times

This stage estimates 3 times duration of activity namely optimistic time (a), optimistic time (b) and most likely time (m).

- Calculate *expected time,* standard deviation and variance on the critical path.

Besides determining the above three times, the next step is to calculate the expected time, standard deviation and variance to determine the expected time by using the formula:

$$te = \frac{a + 4m + b}{6} \quad (1)$$

$$S = \frac{b - a}{6} \quad (2)$$

$$V(te) = S^2 \quad (3)$$

- Calculate the probability of project completion after obtaining the expected time, standard deviation and variance on some critical activities can be continued by calculating the normal z-value. The normal z-value is useful to determine the probability of the completed project within the time frame using normal distribution.

$$Z = \frac{T_d}{t_e} \quad (4)$$

$$Z = \frac{210-204}{11{,}3}$$

$$Z = 0{,}53$$

Td is the duration of the project based on the contract which is 210 days. Based on table of normal distribution of cumulative with value of Z =0,53, the possibility of the project to be completed within 204 days is 79,81 %.

3.4 Project Cost Planning Data

The implementation of a cost project is important things that must be prepared because it is needed as a guide in project implementation.
There are 2 types of cost in a project namely Direct *Cost* and Indirect *Cost.*

1. Direct costs are the cost directly related to the construction of the field. This direct cost is obtained from the multiplication of the volume of job and the unit price of the work. Details of the direct cost of the development graving dock project can be seen in table 1 below:

Table 1. Direct Cost

No.	Activities	Total Cost (IDR)	Code	Duration (Days)
I.	**Preparation Work**			
1	Land Clearing	95.400,00	A	12
2	Measurement & Bouwplank	147.070,00	B	3
3	Keet Direction & Temporary Facility	2.294.857,00	C	2
4	Shop Drawing & Site Engineering	210.000.000,00	D	1
5	Mobilization	471.493.788,00	E	1
6	Damage of Sea Water	242.780.786,00	F	14
II.	**Concrete Sheet Pile Work & Concrete Wall**			
7	Procurement of Concrete Sheet Pile	47.624.227,20	G	3
8	Service for Erection Concrete Sheet Pile	156.121.700,00	H	64
9	Treatment Concrete Sheet Pile	220.199.057,46	I	7
III.	**Excavation Work Ground Area of Graving Dock**			
10	Excavation Work Area Graving Dock	284.680.168,00	J	12

No.	Activities	Total Cost (IDR)	Code	Duration (Days)
11	Excavation Land Disposal Work	353.788.000,00	K	8
12	Land Dredging & Soil Compaction	141.978.166,67	L	15
13	Soil Excavation & Sand Drigging	119.092.924,00	M	10
IV.	**Concrete Work Area of Graving Dock**			
14	Floor/Slab Graving Dock			
	a. Procurement of Diameter Pile 30 cm	26.673.624,00	N	3
	b. Service for Erection	2.508.7009.418,40	O	10
	c. Treatment Pile	143.095.615,49	P	7
	d. Pile Cap	82.390.000,00	Q	8
15	Concrete Floor Work ; Thick 50 cm			
	a. Spawning	6.440.388,00	R	21
	b. Formwork	134.684.220,00	S	4
	c. Casting	198.180.000,00	T	24
	d. Hardener	124.490.321,00	U	21
16	Concrete Wall Work ; Thick 25 cm			
	a. Spawning	2.151.455,00	V	10
	b. Formwork	45.504.360,00	W	7
	c. Casting	66.090.000,00	X	12
	d. Hardener	43.194.107,00	Y	11
V.	**Supporting Facility Work**			
17	Homework Pumps	21.000.000,00	Z	21
18	Making Door of Graving Dock			
	a. Procurement of Plate	1.360.000,00	AA	4
	b. Procurement of Valve	3.129.087,00	AB	14
	c. Fender	110.000.000,00	AC	21
19	Finishing Work	61.363,43	AD	14
20	Demobilisation	471.493.788,00	AE	1
	TOTAL	**6,242,943,891.65**		

2. Indirect costs are costs that are not directly related to production namely overhead cost and unforeseen expenses. Indirect cost is often called distributed cost. Indirect cost may include the cost of any labor that is supportive of the production process. Detail of indirect cost of development graving dock can be seen in table 2 below:

Table 2. Indirect Cost

No	Job Description	Total Cost (IDR)
I. Overhead Cost		
1. Salary of Proyek Staff		
a	Site Manager	33.660.000
b	Civilian Executive	21.420.000
c	Logistik	13.260.000
d	Administration	19.380.000
2. Facility (call, electricity, water, transportation)		30.600.000
II. Unforeseen Expenses (10% of real cost)		624.294.389
Unforeseen Expenses every day		520.245
Total		**742.614.389**

3. Duration of the project construction is one of the key parameter to indicate the success of a project. In order to determine the cost and duration relation therefore it is needed to make S Curve. S curve is used to indicate the progress of the project based on cost and time. Figure 3 shows the S Curve for this project.

No	Activity	Total	Jan	Feb	Mar	Apr	May	Jun	Jul
1	Preparation Work	926.811.901,00							
2	Concrete Sheet Pile Work & Concrete Wall	423.944.984,66							
3	Excavation Work Ground Area of Graving Dock	899.539.258,67							
4	Concrete Work Area of Graving Dock	3.385.603.508,89							
5	Supporting Facilities Work	607.044.283,43							
	TOTAL	6,242,943,891.65	13%	37%	66%	82%	88%	90%	100%

Fig 3. S Curve.

4 Conclusions

Based on the analysis of project scheduling with CPM-PERT method for the development of Graving Dock in Tanjung Intan, Cilacap the following conclusions can be drawn:

1. The cost to develop Graving Dock in Tanjung Intan Cilacap is Rp. 6.985.558.280,65.
2. The critical activities in construction of Graving Dock project in Tanjung Intan is the activities with code A-N-O-P-Q-R-S-T-U-V-W-X-Y-AA-AB-AC-AD-AE and can be completed within 204 days.
3. Based on PERT calculation, it is found that the probability of the shipyard development project on the Tanjung Intan, Cilacap is completed within 204 days is 79,81 %.
4.

References

1. A. Wulandari, *Analisa Peningkatan Kualitas Layanan Jasa Reparasi Kapal di Galangan Kapal Jawa Timur*. 2016, Institut Teknologi Sepuluh Nopember: Surabaya.

2. F.F. Clifford, Erik, W. L., *Manajemen Proyek: Proses Manajerial*. 2007: Andi Publisher, Yogyakarta.
3. M. Aulady, Orleans, C., *Perbandingan Durasi Waktu Proyek Konstruksi Antara Metode Critical Path Method (CPM) dengan Metode Critical Chain Project Management (Studi Kasus: Proyek Pembangunan Apartemen Menara Rungkut).* Jurnal IPTEK Institut Teknologi Adhi Tama, 2016. **20**: p. 14-18.
4. A. N. Andr., *Optimalisasi Penjadwalan Proyek pada Pembangunan Gedung Khusus Laboratorium Stasiun Karantina Ikan Kelas 1 Tanjung Emas Semarang*. 2007, Universitas Negeri Semarang.
5. A. Arinto., *Exploring Bar Chart Method, CPM, PDM, PERT, Line of Balance, and Time Chainage Diagram in Project Scheduling Construction*. 2010, Diponegoro University: Semarang.
6. Silvianita., N., A., Mulyadi, Y., Suntoyo., Chamelia, D. M *Project Scheduling Based on Risk of Gas Transmission Pipe*. in *IOP Conference Series: Earth and Environmental Science*. 2018.
7. Silvianita., F., R., Rosyid, D. M., Suntoyo., Chamelia, D. M. *Fast Missile Boat Project Planning using CPM and What If Analysis Method*. in *IOP Conference Series: Earth and Environmental Science*. 2018.
8. Silvianita, A.N.S., Daniel M Rosyid. *Project Delay Analysis of HRSG*. 2017: IOP Conference Series: Earth and Environmental Science.
9. J. S. Young, J.H.W., Jong Gye Shin, *Research on a Simulation Based Ship Production Support System for Middle Sized Shipbuilding Companies.* International Journal Naval Architec Ocean Engineering, 2009. **1**: p. 70-77.
10. D. Alfianto., *Evaluasi Penjadwalan Proyek dengan Menggunakan Metode CPM dan Analisis Kurva S pada Proyek Studi Detail Desain Pantai Les Tejakula di Kab Buleleng Bali*. 2012, Universitas Pembangunan Nasional Veteran.
11. H. Gitosudarmo., Indriyo, *Sistem Perencanaan dan Pengendalian Produksi*, ed. E. 2. 2000, Yogyakarta: BPFE UGM.

MATEC Web of Conferences **177**, 01029 (2018) https://doi.org/10.1051/matecconf/201817701029
ISOCEEN 2017

Sedimentation Modeling For Development of Kenjeran - Surabaya

Winda Amalia Herdianti[1,*], Suntoyo[2], and Wahyudi[3]
[1]Master Student, Department of Ocean Engineering, Faculty of Marine Technology, Institut Teknologi Sepuluh Nopember, Surabaya, Indonesia (60111)
[2] Associate Professor, Ocean Engineering Department, Faculty of Marine Technology, Institut Teknologi Sepuluh Nopember (ITS), Surabaya, Indonesia (60111).
[3] Assistant Professor, Ocean Engineering Department, Faculty of Marine Technology, Institut Teknologi Sepuluh Nopember (ITS), Surabaya, Indonesia (60111).

Abstract. Surabaya City administratively has a maritime boundary consisting of 11 districts known as Pantai Timur Surabaya (Pamurbaya). One of them is Kenjeran area. The area located in the Development Unit 3 in the Urban Spatial Plan (RTRW) used as a Service and Trade Area and also used as a Tourism Area. Therefore, this study is important to know the shoreline changes and made recommendations in zone management development. Results of this study found that the area of Kenjeran dominated by sedimentation with sedimentation range 0.002-0.015 meters. The largest sedimentation is $2.36x10^6$ m^2 with sedimentation of 0.02 meters. Development of the area in Kenjeran is suitable with the RTRW Surabaya, but due to the large sedimentation, there are some things that must be considered such as the uncertainty of land ownership arise in Kenjeran and ecosystem damage. The solution offered is that the government should restrict the use of arising land and there should be clarification of rules from the government on arising land.
Keyword: Sedimentation, Morphological Changes, Kenjeran-Surabaya

1 Introduction

Coastal area is an area between land and sea, landward covering the land part, either dry or submerged, which is still affected by the nature of the sea such as tides, and sea breeze. Meanwhile towards the sea compromise of the sea that is still influenced by natural processes that occur on land such as sedimentation and fresh water current, or caused by human activities on land such as deforestation and pollution.

The sediment transport process is one of the main proccess in coastal areas. Sediment transport is a displacement of sediment material from a certain place to another. This move is in the form of inflow or outflow. If the outflow is more than the inflow then there will be erosion and vice versa if the outflow is less than the inflow then the accretion process occurs [1, 2, 3]. Generally sediment transport movement is divided into three which is bed load, wash load and suspended load. Wash Load is very fine particles which are transported by the water, but those particles do not exist on the bed. Hence, wash load will not be considered. Suspended load is defined as part of the total transport load which is moving without continous contact with the bed as the result of the agitation of the fluid turbulence. The appearance of ripples will increase

* Corresponding author: windaherdianti@gmail.com

the bed shear stress (flow resistance) [2, 3]. On the other hand, more grains will be suspended due to the flow separation on the lee side of the ripples. Thus the suspended load is related to the total bed shear stress [3]. Bed load is defined which has more or less continuous contact with the bed, the particles are simply rolling, trundle or small jumping at the bottom because of the relatively slow flow rate. Bed load sediment transport modelling under non-linier and irregular waves motion based on bottom shear stress model has been done by [4, 5, 6, 7]. One of the formulas used to calculate the bed load transport by using the DELFT 3D software in the present study is the Engelund Hansen formula [3], as follows

$$S=Sb+Ss,eq=(0.05\alpha\, q^5)/(\sqrt{g}\, C^3\, \Delta^2\, D50) \quad (1)$$

Which is: q = flow velocity (m/s), Δ = relative density, C = Chézy friction coefficient and α = Calibration coeficient.

As a result of changes in sediment's volume is the erosion, sedimentation, and arising land. Kenjeran is one of the coastal areas in Surabaya which become trading and service area, as well as tourism area. Therefore, research on coastal morphology changes in Kenjeran, becomes important because of many functions of spatial and population in this region. The results of this analysis can deliver recommendations in the development of Kenjeran area to improve the welfare of the community and reduce environmental problems.

2 Materials and Method

This research was conducted in three stages: data collection, hydrodynamics and sediment transport modelling using Delft 3D software and spatial study of Surabaya city. The data used include topographic and bathymetry maps, tidal data, sediment properties data, and Surabaya Spatial Plan [8, 9].

Numerical modelling was included the meshing grid, the data input on flow and wave module, and the validation model [3]. This modelling is made to analyze and predict sediment transport movement so that the coastal morphology changes can be seen. The results of this model will be used as a recommendation for the development of Surabaya city in accordance with Surabaya Spatial Plan [8]. In addition, in this study, will also assess the suitability of conditions in the field with the plan by the government in Spatial Planning of Surabaya [8].

3 Result and Discussion

The location in this study was at Surabaya. Geographically located at coordinates 7 ° 14'00 " SL (South Latitude) and 112 ° 47'02 "EL (East Longitude). The area reviewed in this study is Kenjeran-Surabaya with the given following data:

1. The average wind direction: 150^0
2. Average wind speed: 4 knots
3. Average of incoming wave direction: 110^0
4. Significant wave average (Hs): 0.45 meters

5. Average wave period (T): 5.41 s
6. Total suspended solids (TSS) of sediment: 0.0576 Kg / m^3
7. Median sediment grain diameter (D50) : 250 μm
8. Chezy : 7.5
7. Tidal constants given in Table 1.

Table 1. Tidal Constants

Tidal Constants	Amplitude (m)	Phase (deg)
M_8	0.0049	336.15
S_4	0.0046	257.33
MSF	0.0072	354.15
M4	0.004	265.26
MS_4	0.0066	55.15
SK_3	0.0153	08.88
S_2	0.2776	332.96
M_2	0.4199	334.38
K_1	0.4577	288.54
O_1	0.2712	267.4

For a more detailed explanation, the following observation points are selected:

Table 1. Observation point location

No	Observation points	Latitude/Longitude
1.	(233,326)	Latitude :-7.205912^0 Longitude : 112.774690^0
2.	(280,314)	Latitude :-7.217450^0 Longitude : 112.785840^0
3.	(353,288)	Latitude :-7.243352^0 Longitude : 112.801750^0

Based on the result of total transport model was obtained that the data of sedimentation rate in each take at each observation point, where the largest total transport value was occured at the observation point (233,326) as 3.48 x10^{-9} m^3/s/m and the lowest total transport value at the point of observation (280,314) that is 4.62 x10-11 m^3/s/m. The results of sedimentation rate values at each observation point can be seen in Table 3 as follows:

Table 2. Sedimentation rate

Observation point	Bed Load Transport in Average (m^3/s/m)	Suspended Transport in Average (m^3/s/m)	Total Transport in Average (m^3/s/m)
(233,326)	5.11 x10^{-19}	3.48 x10^{-9}	3.48 x10^{-9}
(280,314)	1.01 x10^{-18}	4.62 x10^{-11}	4.62 x10^{-11}
(353,288)	3.38 x10^{-18}	3.66 x10^{-10}	3.66 x10^{-10}

At the observation point (233,326) in Kelurahan Kedung Cowek, there was sedimentation of $7x10^{-5}$ meters, as evidenced by the change of bed level from -5.6915 meters to -5.9615 meters and also the addition of sediment period from 8025 kg / m^2 to 8025.035 kg / m^2 as shown in Figure 1, Figure 2 and Figure 3.

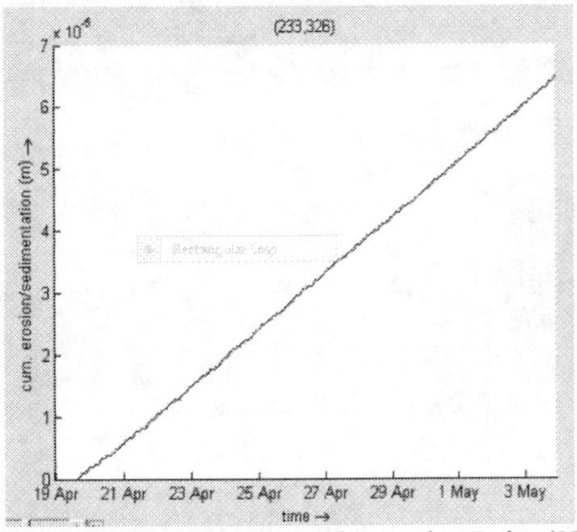

Fig. 1. Cum.Erosion/Sedimentation At Observation Point (233,326)

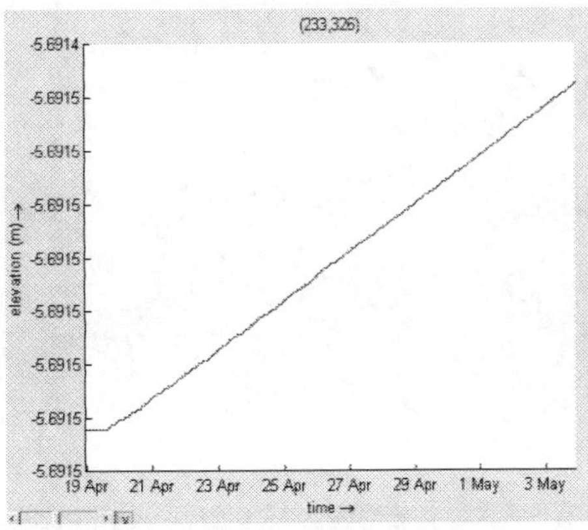

Fig. 2. Bed Level Change At Observation Point (233,326)

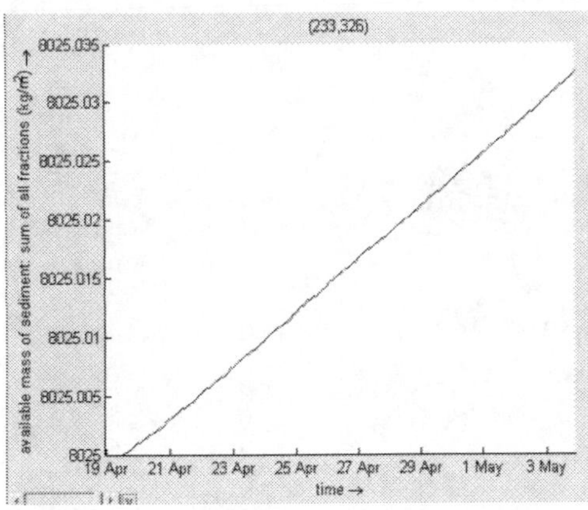

Fig. 3. Available Mass of Sediment At Observation Point (233,326)

At observation point (280,314) in Kenjeran Sub-district sedimentation occurs $1x10^{-3}$ meter, as evidenced by the change of bed level from -2.4337 meters to -2.4328 meters and also the addition of sediment period from 8025 kg / m^2 to 8025.45 kg / m^2 as shown in Figure 4, Figure 5 and Figure 6.

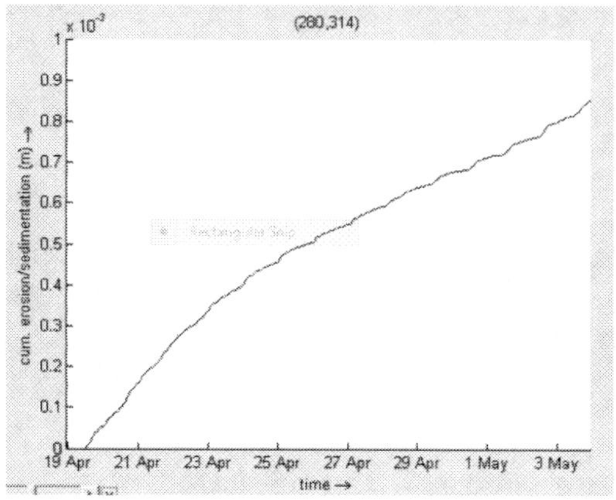

Fig. 4. Cum.Erosion/Sedimentation At Observation Point (280,314)

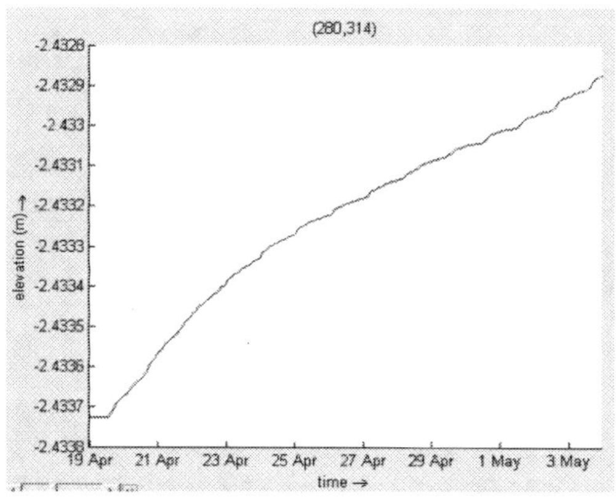

Fig. 5. Bed Level Change At Observation Point (280,314)

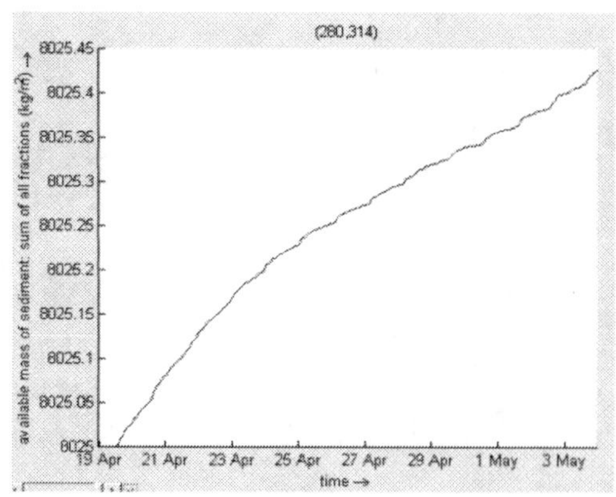

Fig. 6. Available mass of sediment At Observation Point (280,314)

At observation point (353,288) in Sukolilo Sub-district, sedimentation occurs $4x10^{-3}$ meter, as evidenced by the change of bed level from -1.6345 meters to -1.6305 meters and also the addition of sediment period from 8025 kg / m² to 8026.8 kg / m² as shown in Figure 7, Figure 8 and Figure 9.

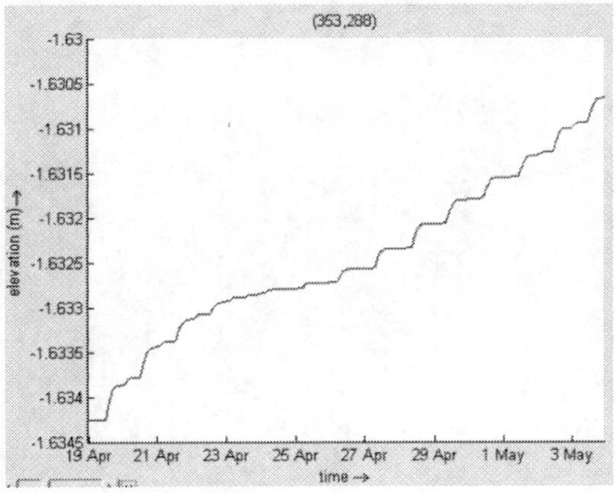

Fig. 7. Cum.Erosion/Sedimentation At Observation Point (353,288)

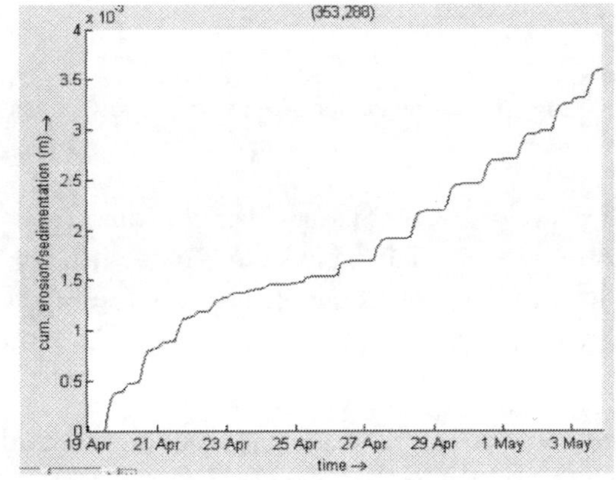

Fig. 8. Bed Level Changes At Observation Point (353,288)

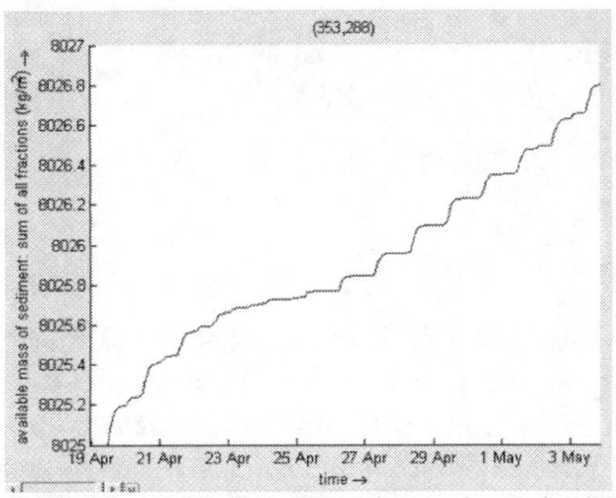

Fig. 9. Available Mass Of Sediment At Observation Point (353,288)

From Tambak Wedi sub-district to Sukolilo sub-district there is extensive sedimentation, which is 2.3x10^6 m^2 area with sedimentation of 0.02 m and 7.5x10^6 m^2 with sedimentation of 0.1 m in the next 15 years based on the sediment transport modelling result as given in Figure 10.

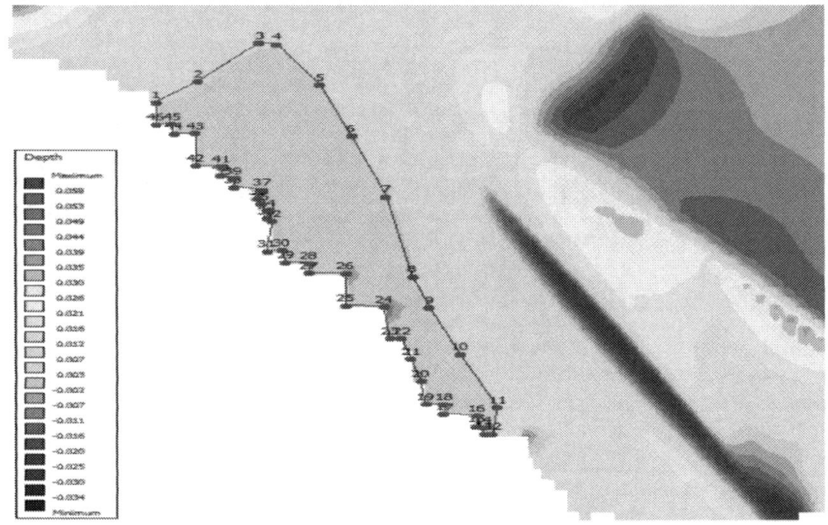

Fig. 10. Sedimentation Accumulation for 15 years

According to the Surabaya City Spatial Plan [8], the area is classified in trade and service areas, and tourism areas. This is in accordance with the conditions in the field that the area is a residential area of fishermen, fishery trading and tourism areas in Old Kenjeran and KenPark. The development of the area by the government is done through:

a. Developing integrated areas of trade and services including provision land for informal traders and built a trading center.
b. Developing an integrated trading area with fulfillment of developer / implementer obligations in the provision of: environmental infrastructure, public utilities, green open space, informal traders and social facilities.
c. Develop and add tourism icons in Old Kenjeran and KenPark.

4 Conclusion

Coastal in Tambak Wedi, Kenjeran and Sukolilo Sub-districts are dominated by sedimetation. The use of this area is in accordance with the Surabaya Urban Spatial Plans, but there are some things that must be done in the future development. Such as limiting the use of arising land in the residential areas of fishermen, conducting the development of environment-based tourism and supervise coastal environment from garbage. So that marine ecosystems are not damaged.

References

1. CERC. *Shore Protection Manual.* US Army Coastal Engineering Research Center. Washington. (1984)
2. Z. Liu. *Sediment Transport. Laboratoriet for Hydraulik og Havnebygning.* Instituttet for Vand, Jord og Miljøteknik Aalborg Universitet (2001)
3. Delft 3D User Manual. *Simulation of multi-dimensional hydrodynamic flows and transport phenomena, including sediments.* Deltares (2014)
4. Suntoyo, H. Tanaka, and A. Sana. Characteristics of turbulent boundary layers over a rough bed under saw-tooth waves and its application to sediment transport. Coastal Engineering, **55** (12), pp. 1102-1112 (2008).
5. Suntoyo, and Tanaka, H. Effect of bed roughness on turbulent boundary layer and net sediment transport under asymmetric waves. Coastal Engineering, **56** (9), pp. 960-969 (2009).
6. Suntoyo, A.H. Fattah, M.Y. Fahmi, T. Rachman and H. Tanaka. Bottom shear stress and bed load sediment transport due to irregular wave motion. ARPN Journal of Engineering and Applied Sciences, **11** (2), pp. 825-829 (2016).
7. Wijaya, M.M., Suntoyo, and Damerianne, H.A. Bottom shear stress and bed load sediment transport formula for modeling the morphological change in the canal water intake. *ARPN Journal of Engineering and Applied Sciences*, **11** (4), pp. 2723-2728 (2016).
8. Review Rencana Tata Ruang Wilayah (RTRW) Kota Surabaya Berdasarkan UU Penataan Ruang No.26/2007 (2009)
9. R. Dahuri. *Pengelolaan Sumber Daya Wilayah Pesisir dan Lautan Secara Terpadu.* Jakarta. PT. Pradnya Paramita. (1996)

MATEC Web of Conferences **177**, 01030 (2018) https://doi.org/10.1051/matecconf/201817701030
ISOCEEN 2017

Study on ultimate strength of ship's hull considering cross section and beam finite element.

Muhammad Zubair Muis Alie[1,*], *Juswan*[1], *Wahyuddin*[1], and *Taufiqur* Rachman[1]

[1]Hasanuddin University, Engineering Faculty, Department of Ocean Engineering, 92172 Gowa, South Sulawesi, Indonesia

Abstract. The objective of the present research is to study the ultimate strength of ship's hull considering cross section and beam finite element under longitudinal bending. The single hull bulk carrier and double hull oil tanker are taken to be analysed. The one-frame space of ship is considered in the calculation. The cross section of ship's hull is divided into element composed plate and stiffened plate. The cross section is assumed to be remained plane and the simply supported is imposed to both side of the cross section. The longitudinal bending moment is applied to the cross section for hogging and sagging condition. The Smith's method is adopted and implemented into the in-house program of the cross section and beam finite element to calculate the ultimate strength of ship's hull. The result of the ultimate strength for hogging and sagging condition obtained by considering the cross section and beam finite element is compared with one another.

1 Introduction

A ship hull for intact condition will against applied loads smaller than the design loads. Cargo loading also acting during voyage, so that the combination between environmental and cargo loading may have an impact to structural damage such grounding, collision, cracks etc. The ultimate strength of ship hull is one of the most important aspects in the ship design criteria. The element of supporting of hull cross section must be adequate against the external forces during voyages. Otherwise, the failure of the ship's hull girder will occur and collapse may happen. Therefore, the analysis of ship's hull girder must be conducted.

The ultimate strength of ship hull girder has been analysed by some researchers. Cui 2016 [1] focused on the ultimate strength of typical bottom structure in container ships under both longitudinal and transversal loads in corrosive environment. Muis Alie, M.Z et al 2017 [2] assessed the ultimate hull girder strength of Ro-Ro ship after damage. The cross section of Ro-Ro ship is considered to be analysed. Muis Alie, M.Z et al 2016 [3] analysed the hull girder ultimate strength of asymmetrically damaged ships using Finite Element Method. The collision damage was modelled by removing the plate and stiffened plate elements. Muis Alie, M.Z et al 2014 [4] analysed the residual strength of ship hull girder with bottom damage. The nonlinear finite element method was used and the fully cross section was considered in the calculation. Shu, Z 2012 [5] investigated the ultimate strength of a capsize bulk carrier hull girder under combined global and local load in the hogging and alternate hold loading (AHL) condition using nonlinear finite element (FE) analysis with ABAQUS. Amlashi, H.K.K 2008 [6] contributed to establish rational ultimate longitudinal strength criteria for the hull girder under combined loading. An important issue

* Corresponding author: zubair.m@eng.unhas.ac.id

was the significant double bottom bending in empty holds in AHL due to combined global hull girder bending moment and local loads. Paik, J.K 2001 [7] investigated the ultimate strength characteristics of ship hulls with large hatch opening under torsion. Axial (warping) as well as shear stress is normally developed for thin-walled beams with open cross sections subjected to torsion.

In the present paper, the ultimate strength of ship's hull considering cross section and beam finite element under longitudinal vertical bending moment is studied. The single hull bulk carrier and double hull oil tanker are taken to be analysed. The one-frame space of ship is considered in the calculation. The cross section of ship's hull is divided into elements composed plate and stiffened plate. The cross section is assumed to be remained plane and the simply supported is imposed to both side of the cross section. The longitudinal bending moment is applied to the cross section for hogging and sagging condition. The Smith's method is adopted and implemented into the in-house program of the cross section and beam finite element to calculate the ultimate strength of ship's hull. The result of the ultimate strength for hogging and sagging condition obtained by considering the cross section and beam finite element is compared with one another.

2 Cross Section Approach

The ultimate strength of ship hull is performed using the Smith's method considering the cross sectional approach. The in-house program code HULLST developed by Yao and Nikolov [8] is used with some modification in subroutine program. The moment-curvature relationship for the ultimate hull girder strength calculation is expressed by,

$$\begin{Bmatrix} \Delta P \\ \Delta M_H \\ \Delta M_V \end{Bmatrix} = \begin{bmatrix} \overline{D}_{AA} & \overline{D}_{AV} & \overline{D}_{AH} \\ \overline{D}_{HA} & \overline{D}_{HH} & \overline{D}_{HV} \\ \overline{D}_{VA} & \overline{D}_{VH} & \overline{D}_{VV} \end{bmatrix} \begin{Bmatrix} \Delta \varepsilon_0 \\ \Delta \varphi_H \\ \Delta \varphi_V \end{Bmatrix} \quad (1)$$

Where

$$\begin{aligned} \overline{D}_{AA} &= \sum_{i=1}^{N} D_i A_i \\ \overline{D}_{HH} &= \sum_{i=1}^{N} D_i y_i^2 A_i \\ \overline{D}_{VV} &= \sum_{i=1}^{N} D_i z_i^2 A_i \\ \overline{D}_{AH} = \overline{D}_{HA} &= \sum_{i=1}^{N} D_i y_i A_i \\ \overline{D}_{AV} = \overline{D}_{VA} &= \sum_{i=1}^{N} D_i z_i A_i \\ \overline{D}_{HV} = \overline{D}_{VH} &= \sum_{i=1}^{N} D_i y_i z_i A_i \end{aligned} \quad (2)$$

To obtain the ultimate hull girder moment capacity, the above expression can be simply expressed by,

$$\begin{Bmatrix} 0 \\ \Delta M_H \\ \Delta M_V \end{Bmatrix} = \begin{bmatrix} \overline{D}_{AA} & 0 & 0 \\ 0 & \overline{D}_{HH} & \overline{D}_{HV} \\ 0 & \overline{D}_{VH} & \overline{D}_{VV} \end{bmatrix} \begin{Bmatrix} \Delta\varepsilon_0 \\ \Delta\phi_H \\ \Delta\phi_V \end{Bmatrix} \tag{3}$$

Where

$$\begin{aligned} \overline{D}_{AA} &= \sum_{i=1}^{N} D_i A_i \\ \overline{D}_{HH} &= \sum_{i=1}^{N} D_i (y_i - y_G)^2 A_i \\ \overline{D}_{HV} = \overline{D}_{VH} &= \sum_{i=1}^{N} D_i (y_i - y_G)(z_i - z_G) A_i \\ \overline{D}_{VV} &= \sum_{i=1}^{N} D_i (z_i - z_G)^2 A_i \end{aligned} \tag{4}$$

3 Beam Finite Element Method

The other approach to calculate the ultimate hull girder strength of ship hull is beam finite element method. Basically, the calculation based on the fundamental of thin-walled beam theory by considering warping effect. The Smith's method is also adopted and implemented into the in-house program with some of the subroutines are combined with HULLST to generate the cross section becomes beam finite element. The formulation of the beam finite element method to analyze the ultimate hull girder strength may be calculated by,

$$\begin{Bmatrix} \sigma_z \\ \tau_{sz} \end{Bmatrix} = \begin{bmatrix} d_{11} & d_{12} \\ d_{21} & d_{22} \end{bmatrix} \begin{Bmatrix} \varepsilon_z \\ \gamma_{sz} \end{Bmatrix} \tag{5}$$

$$\begin{pmatrix} \varepsilon_z \\ \gamma_{sz} \end{pmatrix} = \begin{bmatrix} [B_1] & -x[B_2] & -y[B_3] & \omega_{ns}[B_2] \\ 0 & 0 & 0 & g(s)[B_3] \end{bmatrix} \{d\} \tag{6}$$

Where

$$\begin{aligned} [B_1] &= \frac{d}{dz}[A_L] \\ [B_2] &= \frac{d^2}{dz^2}[A_C] \\ [B_3] &= \frac{d}{dz}[A_C] \\ g(s) &= \frac{\partial \omega_{ns}}{\partial s} - (y - y_s)\frac{\partial x}{\partial s} + (x - x_s)\frac{\partial y}{\partial s} \end{aligned} \tag{7}$$

$$[K]=\int_V \begin{bmatrix} d_{11}[B_1] & -xd_{11}[B_{12}] & -yd_{11}[B_{12}] & \omega_{ns}d_{11}[B_{12}] \\ -xd_{11}[B_{21}] & x^2d_{11}[B_2] & xyd_{11}[B_2] & -x\omega_{ns}d_{11}[B_2] \\ -yd_{11}[B_{21}] & xyd_{11}[B_2] & y^2d_{11}[B_2] & -y\omega_{ns}d_{11}[B_2] \\ \omega d_{11}[B_{21}] & -x\omega_{ns}d_{11}[B_2] & -y\omega_{ns}d_{11}[B_2] & \omega_{ns}^2d_{11}[B_2]+g^2d_{22}[B_3] \end{bmatrix} \quad (8)$$

4 Results and Discussions

The Smith's method is adopted for the calculation of the ultimate strength of ship's hull under longitudinal bending moment. The one-frame space of ship is considered to be analyzed. The simpy supported of boundary condition is applied to both side of the cross section and beam finite finite element. The welding residual stress, initial imperfection and crack are not considered in the analysis. To know the ultimate strength of ship hull by considering the cross sectional approach and beam finite element method, the intact condition is taken. The sample of the ships are the single hull bulk carrier consists of Ship B1 and Ship B2 as shown in fig. 1, and double hull oil tanker in fig. 2. Ship B1 and Ship B2 for single hull bulk carrier have different dimension also the number of element. The boundary condition is assumed to be simply supported at both side for cross sectional approach and beam finite element method as illustrated in fig. 3.

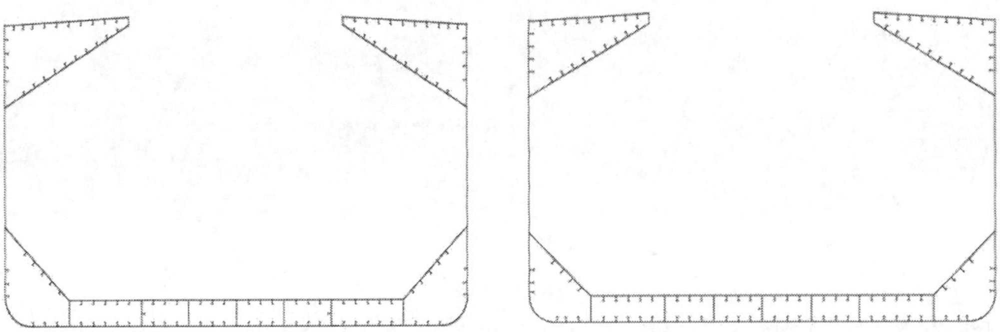

(a) Ship B1 **(b)** Ship B2

Fig. 1. Single Hull Bulk Carrier

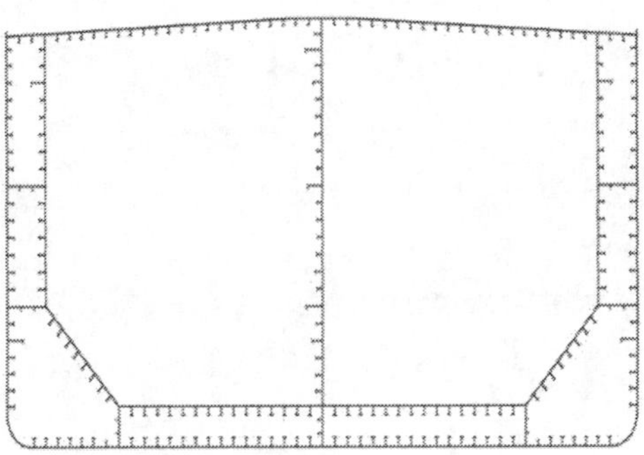

Fig. 2. Double Hull Oil Tanker

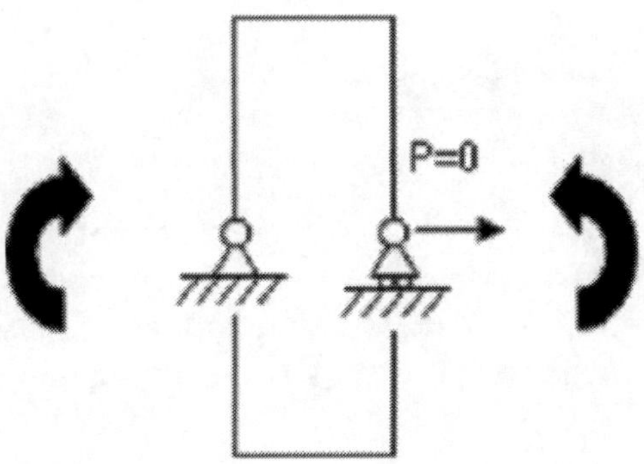

Fig. 3. Boundary Condition

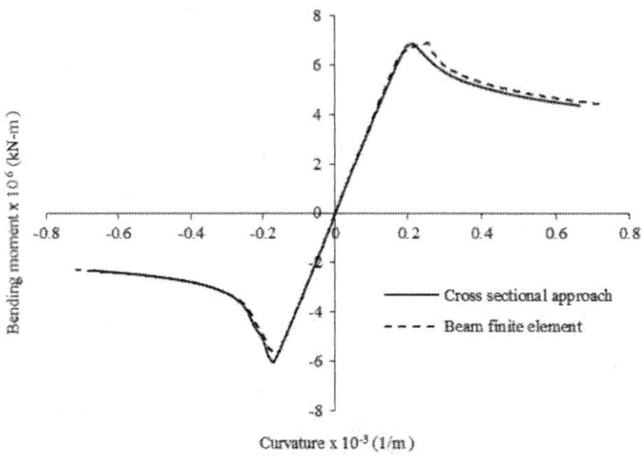

(a) Ship B1

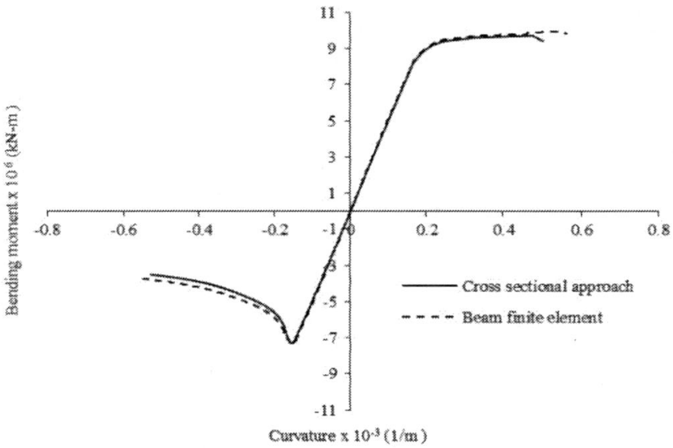

(b) Ship B2

Fig. 4. Moment-curvature relationship for single hull bulk carrier

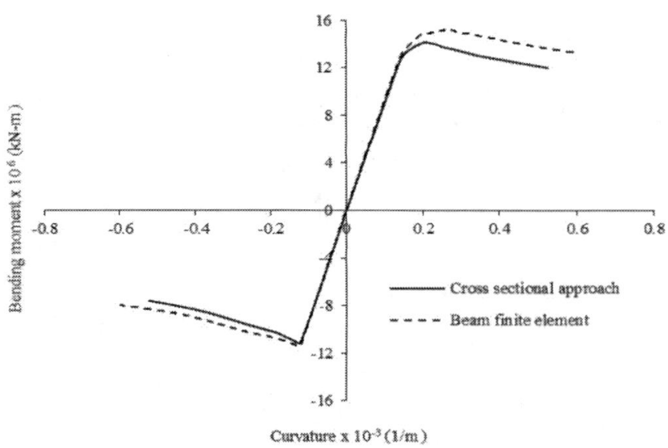

Fig. 5. Moment-curvature relationship for double hull oil tanker

The moment-curvature relationship for ship B1 and ship B2 is expressed in fig. 4. The solid line describe the ultimate strength obtained by cross sectional approach and dot one produced by beam finite element. It is observed that the ultimate strength obtained by cross sectional approach and beam finite element is almost identic both ship B1 and ship B2 in hogging and sagging condition.

Fig. 5 shows the moment-curvature relationship for double hull oil tanker. It is observed that the ultimate strength in hogging condition is larger than sagging condition. Probably, this may due to the element component, element type and number of element of the deck part is quite different with bottom part since the deck is tension and bottom is compression.

5 Results and Discussions

The ultimate strength obtained by cross sectional approach and beam finite element is in good aggreement both ship B1 and ship B2 in hogging and sagging condition. For double hull oil tanker, the ultimate strength in hogging condition is larger than sagging condition. Probably, this may due to the element component, element type and number of element of the deck part is quite different with bottom part since the deck and bottom parts are in tension and compression, respectively.

References

1. Cui, J. Wang, D. Ma, N. Proc. 27th Int Offshore and Polar Eng Conf, ISOPE, 990–997, 2016.
2. Muis Alie, M.Z, Sitepu G, and Latumahina, S.I, Proc. 27th Int Offshore and Polar Eng Conf, ISOPE, 913–919, 2017.
3. Muis Alie, M.Z, Sitepu G, Juswan, Wahyuddin, Nugraha, A.M and Alamsyah," Proc. 35th Int Conf. on Ocean, Offshore and Arctic Engineering, OMAE, 2016.
4. Muis Alie, M.Z, Proc. 3rd Int Conf. on Ship and Offshore Technology, ICSOT, 1-4, 2014.
5. Shu, Z and Moan, T, J. Mar Sci. Technol, 94-113, 2012
6. Amlashi H, K.K and Moan, T, Marine Structures, 327-352, 2008.
7. Paik, J.K, Thayamballi, A.K, Pedersen, P.T and Park, Y.K, Ocean Engineering, 1097-1133, 2001.
8. Yao, T and Nikolov, P.I, J. Soc. Naval Arch of Japan, 437-446, 1992.

AUTHOR INDEX

EDP Sciences – Web of Conferences
17, Avenue du Hoggar
Parc d'Activité de Courtabouf
BP 112
F-91944 Les Ulis Cedex A
France

ISBN 978-1-5108-6600-3

CURRAN ASSOCIATES INC.
proceedings.com